Lecture Notes in Mathematics 1867

James Sneyd (Ed.)

Tutorials in Mathematical Biosciences II

Mathematical Modeling of Calcium Dynamics and Signal Transduction

With Contributions by:

R. Bertram · J.L. Greenstein · R. Hinch
E. Pate · J. Reisert · M.J. Sanderson
T.R. Shannon · J. Sneyd · R.L. Wilson

Editor

James Sneyd

Department of Mathematics
University of Auckland
Private Bag 92019
Auckland 1020, New Zealand

E-mail: sneyd@math.auckland.ac.nz

Cover Figure: Calcium signalling: dynamics, homeostasis and remodeling
© 2003 Nature Publishing Group

Library of Congress Control Number: 2004109594

Mathematics Subject Classification (2000): 92C37, 92C30, 92C05

ISSN print edition: 0075-8434
ISSN electronic edition: 1617-9692
ISBN-10 3-540-25439-0 Springer Berlin Heidelberg New York
ISBN-13 978-3-540-25439-6 Springer Berlin Heidelberg New York

DOI 10.1007/b107088

Springer is a part of Springer Science+Business Media
springeronline.com

Printed in The Netherlands

Typesetting by the authors and Techbooks using a Springer LaTeX package
Cover design: *design & produktion* GmbH, Heidelberg

Printed on acid-free paper SPIN: 11406501 41/Techbooks 5 4 3 2 1 0

Preface

This is the second volume in the series "Lectures in Mathematical Biosciences". These lectures are based on material which was presented in tutorials or developed by visitors and postdoctoral fellows of the Mathematical Bioscience Institute (MBI), at The Ohio State University. The aim of this series is to introduce graduate students and researchers with just a little background in either mathematics or biology to mathematical modeling of biological processes. The present volume is devoted to Mathematical Modeling of Calcium Dynamics and Signal Transduction, which was the focus of the MBI program in the winter of 2004; documentation of that program, including streaming videos of the workshops, can be found on the website http://mbi.osu.edu.

This volume was organized and edited by James Sneyd. Sneyd is a world leader in mathematical physiology and biology, and has been working extensively on modeling biological processes of signal transduction induced by calcium oscillations.

Some of the chapters describe mathematical models of calcium dynamics as they occur in signal transduction. However, more attention is given in this volume to the underlying physiology, since, as Sneyd says, "Mathematical physiology is not possible without the physiology."

I wish to express my thanks to the contributors, all of whom served also as tutorial lecturers at the MBI. Special thanks are due to James Sneyd, who took it upon himself to organize this volume. I hope this volume will serve as a useful introduction to those who are interested in learning about mathematical physiology, and maybe even participating in this exciting field of research.

April, 2005 *Avner Friedman*

Contents

Introduction

The question of how cells respond to their environment and coordinate their behavior with that of other cells is one that can naturally be studied using mathematical models. In order for cells to communicate with each other or with the outside world they have developed a large number of transduction mechanisms, whereby extracellular signals can be translated into intracellular signals, or a signal of one type can be changed into a signal of another type. For instance, muscle cells change an electrical signal into a force; photoreceptors change a light signal into an electrical signal; neurosecretory cells change an electrical signal into a hormonal signal; while in many cell types binding of a neurotransmitter or a hormone leads to oscillations in the concentration of intracellular free calcium, oscillations which control a variety of intracellular processes, including secretion, gene expression, cell movement, or wound repair. For instance, in muscle cells, the release of calcium from the sarcoplasmic reticulum controls muscle contraction, while in olfactory neurons and photoreceptors calcium forms an important negative feedback loop that controls adaptation. In neurosecretory cells, oscillations of the cytoplasmic calcium concentration lead to hormone secretion, while in neurons, calcium is not only crucial for synaptic communication, it is also an important modulator of synaptic plasticity.

In this volume we present a number of examples of signal transduction, showing how physiology and mathematical modeling interact to give a detailed quantitative understanding of such complex phenomena. Because of the widespread importance of calcium, all the chapters here will necessarily include discussion of calcium dynamics. Thus, we begin by showing how mathematical models of calcium dynamics can be constructed and analyzed. This is followed by descriptions of excitation-contraction coupling (i.e., how calcium forms a link between the muscle action potential and the contractile proteins), how the muscle proteins themselves work, and, finally, chapters on the physiology and modeling of olfactory neurons and of neuronal synapses.

Although this volume is part of a series of Tutorials in Mathematics, readers will soon notice that much of what is presented here is not mathematics at all, but physiology. This is entirely deliberate. It cannot be emphasized too strongly that mathematical physiology is not possible without the physiology. Without a detailed understanding of the physiology of the system under discussion it is simply not possible to say anything very interesting about it, no matter how clever is the mathematics used. Three of the authors here (Sanderson, Shannon and Reisert) are experimental physiologists, each of whom works closely with mathematical modelers to incorporate sophisticated modeling techniques into their research. The other four authors are primarily modelers, but ones that work closely with physiologists, or even, in some cases, do some experimental work themselves.

It is to be hoped that this combination of physiology and mathematics in what is, primarily, a volume for mathematicians, will be of use to all those who are interested in learning how modeling is done, or maybe even participating themselves in this most satisfying of endeavors.

April, 2005 *James Sneyd*

Basic Concepts of Ca^{2+} Signaling in Cells and Tissues

M.J. Sanderson

Department of Physiology, University of Massachusetts Medical School, Worcester, MA 01655

1 Introduction

Living tissues are complex organizations of individual cells and to perform their specific functions the activity of each cell within the tissue must be regulated in a coordinated manner. The mechanisms through which this regulation occurs can be equally complex, but a common way to exert control is via neural transmission or hormonal stimulation. Irrespective of the organization of the extracellular control system, the regulatory signals need to be translated into an intracellular messenger that can modulate the cellular processes. Again, there are a variety of intracellular messengers that achieve this aim, including cAMP, cGMP and NO, but here we focus on the calcium ion as the internal messenger. The objective of this article is to provide an overview of the basic mechanisms of how Ca^{2+} serves as a signaling messenger. For greater detail, the reader must refer to the many extensive reviews (for example, Berridge et al., 2003; Berridge et al., 2002). The details of the individual mechanisms are extremely important since they can confer specificity on the signaling model. As a result, model simulations of Ca^{2+} signaling are most useful when the model is designed for a specific cell type and sufficient experimental detail can be incorporated.

2 Ca^{2+} Stores and Pumps

A key characteristic determining how Ca^{2+} serves as a signaling molecule is that, unlike organic messengers (i.e. cAMP, NO), Ca^{2+} ions cannot be created or metabolized. Consequently, cell signaling with Ca^{2+} ions requires a strategy of accumulation or storage coupled with Ca^{2+} release. The cell exploits two major Ca^{2+} stores, the external environment that contains a virtually infinite supply of Ca^{2+} at a concentration of about 1–2 mM and an internal store, usually contained within the endoplasmic (ER) or muscle-equivalent sarcoplasmic (SR) reticulum (Petersen et al., 2001). The storage capacity of the E/SR is enhanced by the presence of Ca^{2+} buffering proteins (e.g. calsequestrin and calretinin) that can bind large quantities of Ca^{2+}. The free Ca^{2+} concentration in the ER is in the 10–100 µM range. Mitochondria can also serve as a Ca^{2+} reservoir, taking up Ca^{2+} in times of excess and releasing Ca^{2+} when cytosolic Ca^{2+} falls. Another, but less-well defined, secondary Ca^{2+} store is the lysosome-related store (Fig. 1).

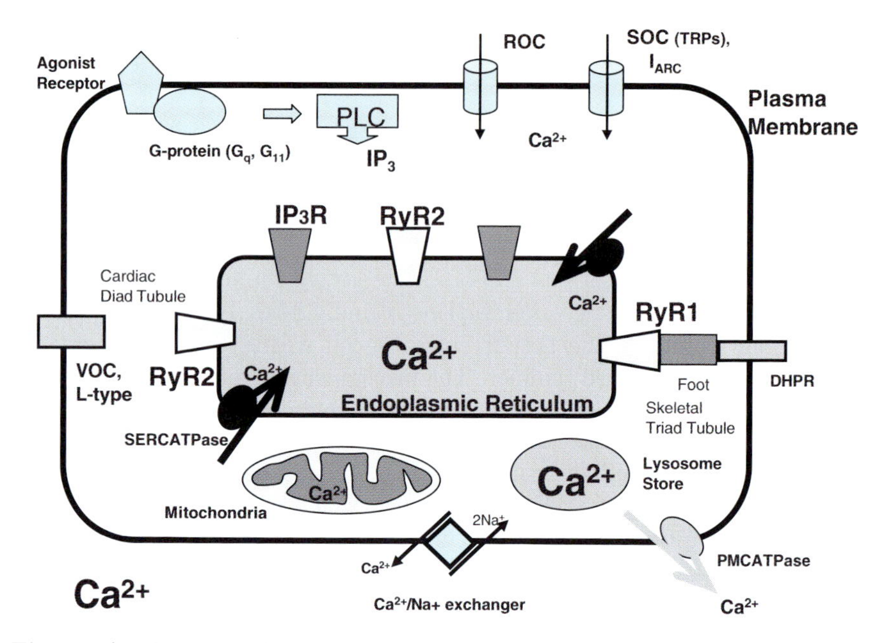

Fig. 1. A schematic of a generalized cell illustrating the fundamental elements of the Ca^{2+} signaling machinery. Ca^{2+} stores exist in the extracellular medium, the endoplasmic reticulum, the mitochondria and the lysosome. Ca^{2+} is moved to these stores by a Ca^{2+}/Na^{+} exchanger, plasma membrane Ca^{2+} pumps and SERCA pumps. Ca^{2+} can enter the cytoplasm via receptor-operated channels (ROC), store-operated channels (SOC), voltage-operated channels (VOC), ryanodine receptors (RyR) and inositol trisphosphate receptors (IP_3R). Agonist stimulation acts via G-proteins to stimulate phospholipase C (PLC) to produce IP_3. RyR receptors are coupled with VOCs in cardiac muscle or with the DHPR in skeletal muscle

To maintain the necessary electrochemical gradient that drives Ca^{2+} release from these stores, the cell membranes separating the stores must be relatively impermeable to Ca^{2+}. In addition, the cell must be able to lower the cytoplasmic Ca^{2+} concentration and re-load the stores to turn off the signal to reset the system. This is achieved by energy-dependent Ca^{2+} pumps (Ca^{2+}ATPases) or Ca^{2+} exchangers located in the cell membranes. There are 2 basic forms of Ca^{2+} ATPase; outer plasma cell membrane ATPases (PCMAs) and sacroplasmic/endoplasmic reticulum calcium ATPase (SERCA) pumps

(Strehler and Treiman, 2004). There is a variety of isoforms of both types of pump. The function of SERCA pumps can be inhibited by pharmacological agents (i.e. thapsigargin) and this approach is frequently used to assess the relative contribution of the internal store to the Ca^{2+} signals. For math modeling, the rates at which these two pumps operate must be considered along with the leakage flux of Ca^{2+} between the two stores (Fig. 1).

Because the cell membranes separating the Ca^{2+} stores from the cytosol are, for the most part, impermeable to Ca^{2+}, the rapid movement of Ca^{2+} across these membranes that is required to mediate a control signal requires a release mechanism (Bootman et al., 2002). As a result, the cell has evolved a spectrum of Ca^{2+}-permeable channels. The distribution of these channels within different cell types can vary considerably to determine a specific signaling mechanism.

3 Ca^{2+} Release Channels

Ca^{2+} release channels exist in a variety of forms and, perhaps, the most widely known are the voltage-operated channels (VOCs). These channels are usually found in the plasma membrane of excitable cells where, in response to membrane depolarization switch from a closed to an open state. The permeability of the channel is determined by the molecular characteristics of the channel pore. There are 4 major types of Ca^{2+}-permeable VOCs (P, N, L, Q-type) and their characteristics vary with respect to opening duration and conductance. Ca^{2+}-permeable channels also exist as stretch or mechanically-operated channels.

A second class of Ca^{2+} release channel is the receptor-operated channels (ROCs). These channels are also normally on the plasma membrane and open upon binding of an agonist to the channel. The third class of plasma membrane Ca^{2+} channels is store-operated channels (SOCs). These channels are believed to open in response to the emptying of the internal Ca^{2+} stores and their primary function is the replenishment of the internal Ca^{2+} stores although a steady influx of Ca^{2+} via these channels may itself serve as a biological signal (Nilius and Voets, 2004; Putney, 2004). However, the identity of these channels and the mechanism by which they open remains controversial. While specific transient receptor potential (TRP) channels are thought to be synonymous with SOCs, an alternative Ca^{2+} influx pathway also appears to occur via an arachidonic acid activated channel (I_{ARC}) (Shuttleworth, 2004).

In terms of Ca^{2+} release from internal stores, the two most important channels (although they are called receptors) are the ryanodine receptor (RyR) and the inositol trisphosphate receptor (IP_3R) (Meissner, 2004; Taylor et al., 2004). These channels/receptors must bind Ca^{2+} and/or IP_3 in order to open and therefore may be considered a specialized form of ROC (Fig. 1). Another agonist of the RyR is cADP-ribose. NAADP can also release Ca^{2+} from other internal stores but the receptor mediating this effect is unknown (Lee, 2004).

4 Signaling by Ca^{2+} Transients

Many cell responses can be mediated through a single or transient elevation in intracellular Ca^{2+}. A good example of cell signaling that couples an external stimulus to an internal release of Ca^{2+} is the contraction of cardiac muscle. Cardiac muscle contraction is initially stimulated by depolarization of the cell membrane that activates membrane-bound VOCs (L-type channels). This allows a small amount of Ca^{2+} to enter the cell (Fig. 2). However, muscle contraction must last considerably longer than the membrane depolarization and the Ca^{2+} signal must also extend deeper into the cell to fully activate the contractile filaments. This requires amplification of the Ca^{2+} signal and is achieved by the distribution of RyRs on the SR.

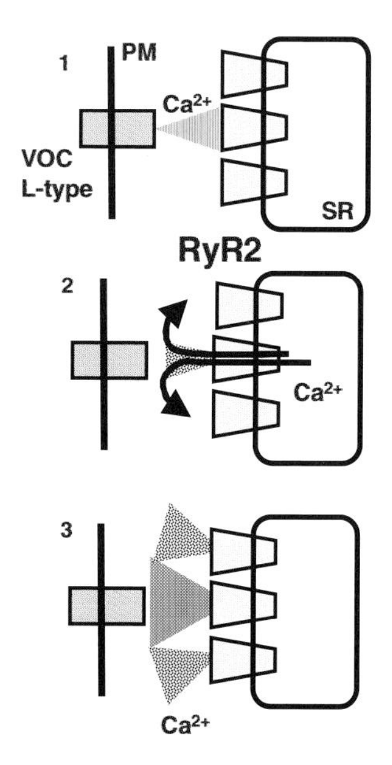

Fig. 2. The sequence of events stimulating Ca^{2+} release for contraction of cardiac muscle. (**1**) Membrane depolarization opens VOCs in the plasma membrane to initiate an influx of Ca^{2+}. (**2**) The RyR in the proximity of the VOC is induced to open by the exposure to Ca^{2+}. (**3**) The Ca^{2+} released by one RyR diffuses to adjacent RyRs and through the process of CICR induces the further release of Ca^{2+}

Ryanodine receptors: There are three basic types of RyRs, type 1, 2 and 3. The RyR of cardiac muscle is type 2 and similar to RyR receptors found in non-excitable cells. The RyR2 is a large macromolecule constructed from 4 subunits (each about 560 kDa). The membrane spanning loops of the protein subunits form a central pore for Ca^{2+} flux, but a large portion of the molecule does not reside within the membrane and projects above the SR (Fig. 1). It is in this part of the molecule where Ca^{2+} and a variety of other proteins or agonists such as calmodulin, FK506, ATP or cADP-ribose can bind to influence the activity of the RyR. However, the primary event responsible for the opening of the RyR2 is the binding of Ca^{2+}. Not only is the RyR tuned to respond to Ca^{2+}, but in cardiac muscle its location is also optimized, in a dual membrane structure called a diad, to be in the vicinity of the L-type Ca^{2+} channels in order to maximize its exposure to the small amount of Ca^{2+} influx during signaling. Thus a small Ca^{2+} signal is amplified into a larger Ca^{2+} pulse by opening the RyR (Fig. 2). However, the amplification does not stop here. The Ca^{2+} released by the RyR itself has a positive-feed back on RyRs in the immediate vicinity to induce them to also open and release Ca^{2+}. This will result in a local high concentration of Ca^{2+} around the RyR that is sufficient to stimulate contraction. At much higher concentrations, Ca^{2+} can also bind to additional low affinity sites of the RyR. The effect of this is to close the RyR and thereby limit the local amplification process. This process of Ca^{2+} signal amplification and propagation is used by a variety of cells and is commonly referred to as Ca^{2+}-induced Ca^{2+} release (CICR). Although negative Ca^{2+} feedback is essentially controlling the Ca^{2+} transient, the exact magnitude of the Ca^{2+} transient can be modified in a variety of ways and this is used to vary the strength of the muscle contraction. For example, phosphorylation of the RyR or the degree of SR loading with Ca^{2+} can alter the opening behavior of the RyR.

The type 1 RyR, found predominately in skeletal muscle performs a similar function to the RyR2 but instead of initially relying on Ca^{2+} diffusion from a membrane Ca^{2+} channel, the cytosolic portion of the RyR is modified into a "foot" region that makes a direct link with a membrane voltage sensor – the dihydropyridine receptor (DHPR) (Fig. 1). With this modification, depolarization is directly translated into RyR channel opening. However, further signal amplification can still occur via CICR through other RyRs.

5 Cell Signaling by Ca^{2+} Oscillations

In contrast to a transient response of contracting muscle, other cells may require a sustained response, i.e. smooth muscle cells that must maintain tone over long periods. In addition, many cells are non-excitable and can only respond to diffuse hormonal stimuli rather than to a discrete depolarization pulse. An obvious way to maintain a sustained response would be to sustain an elevated intracellular Ca^{2+} concentration. However, prolonged periods of elevated levels of Ca^{2+} are frequently toxic to cells. In addition, in view of the normal variability associated with a Ca^{2+} signal, it is difficult to determine the magnitude of a Ca^{2+} elevation.

The solution to these problems appears to be the generation of repetitive Ca^{2+} signals or Ca^{2+} oscillations within a cell. The principal internal element responsible for generating this type of signaling is the IP_3R of the ER. There are 3 major isoforms of the IP_3R and these have significant structural homology with the RyR. Consequently, the IP_3R is a large tetrameric molecule imbedded in the ER/SR. The majority of the molecule extends into the cytoplasm and the distal end of the molecule can bind IP_3. Like the RyR, the IP_3R can bind Ca^{2+} at a high (activating) and low (inactivating) affinity sites on each subunit.

The initial step in generating Ca^{2+} oscillations is the translation of the external signal into the intracellular production of the second messenger IP_3. Usually IP_3 production results from the activation of the membrane-bound enzyme phospholipase C (PLCβ), via G-protein receptors, to hydrolyze phosphatidylinositol-4, 5-bisphosphate into IP_3 and diacylglycerol (DAG). There are numerous receptor types that are coupled via G proteins to PLC. In addition, other isofoms of PLC can be activated by tyrosine kinase–coupled receptors (PLCγ) or Ca^{2+} (PLCε). The activation process can also occur by providing the PLC itself as is the case at fertilization. The sperm donates PLCζ to the ovum for subsequent rounds of Ca^{2+} signaling (Swann et al., 2004).

Under starting conditions of low levels of Ca^{2+}, the binding of IP_3 to the IP_3R results in its opening (Fig. 3). The release of Ca^{2+} from the ER then serves to further activate the IP_3R by increasing its open probability. This leads to further increases in Ca^{2+} concentration in the vicinity of the receptor that ultimately stimulates the binding of Ca^{2+} to a low affinity site which in turn reduces the IP_3R open probability. In summary, the open probability of the IP_3R is a function of Ca^{2+} concentration and is a bell-shaped function (Fig. 3). At higher IP_3 concentrations, this function is shifted upwards and to the right. In the presence of a constant concentration of IP_3, repetitive increases in Ca^{2+} are generated by sequential positive and negative feedback control influenced by Ca^{2+}. Models of the Ca^{2+} signaling usually include a sub-model of the IP_3R based on 4 subunits which require the binding 1 IP_3 molecule and 2 Ca^{2+} ions.

Again, the characteristics of the Ca^{2+} oscillations are determined by the details of the system in different cells. For example, the sensitivity of system can be determined by IP_3R type or the level of phosphorylation of the IP_3R. However, the most important characteristic of the Ca^{2+} oscillations is the frequency at which they occur. This is determined by the concentration of IP_3 which in turn is a function of PLC activity, the regulatory G-proteins and agonist stimulation.

5.1 FM Regulation of Cell Activity

The advantage of signaling by Ca^{2+} oscillations is that the signal can be encoded in the frequency of the oscillations (frequency-modulated, FM) rather than in the magnitude of the Ca^{2+} signal. Indeed, the magnitude of successive Ca^{2+} oscillations is very similar. The only important requirement is that each oscillation should be easily distinguished from the background noise. However, a pre-requisite for FM signaling is a detection system that can decode

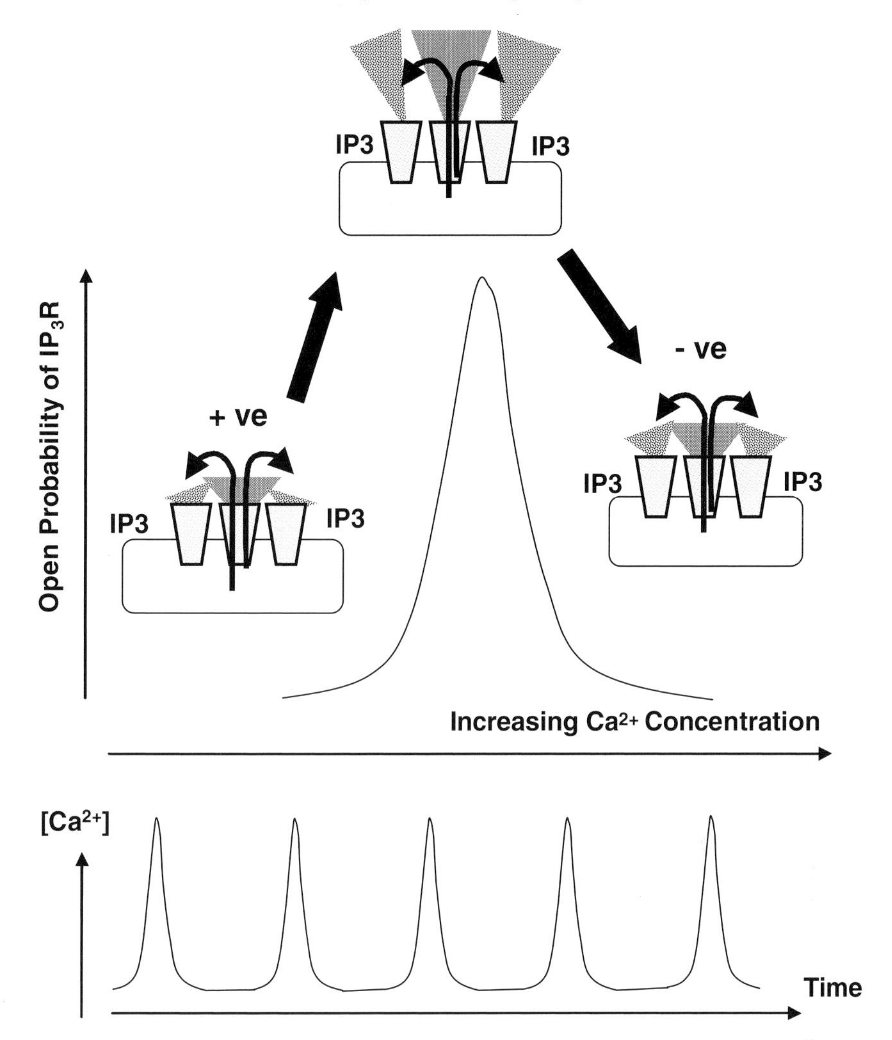

Fig. 3. The open probability of the IP_3R. At low concentrations of Ca^{2+}, the binding of IP_3 to the IP_3R stimulates receptor opening and Ca^{2+} release. The Ca^{2+} released exerts a positive (+) feedback to increase the open probability of the IP_3R and this increases the amount of Ca^{2+} released. The increased Ca^{2+} concentration exerts a negative (−) feedback on the open probability of the IP_3R, resulting in the closure of the receptor. This allows the Ca^{2+} pumps to reduce the cytosolic Ca^{2+} concentration. Repetitive cycles of opening and closing of the IP_3R result in a series of Ca^{2+} oscillations (bottom) in the presence of a constant concentration of IP_3

the frequency information. The basic design of a molecular detector is that the activation reaction driven by a single Ca^{2+} oscillation is substantially faster than the inactivation process that occurs during the period of reduced Ca^{2+}. Generally, this takes the form of phosphorylation driven by Ca^{2+}-calmodulin-activated protein kinases and de-phosphorylation driven by phosphatases. A good example is Cam-Kinase II (Hudmon et al., 2002). The activity of this enzyme increases with phosphorylation of its multiple self-phosphorylatable sites and successive Ca^{2+} oscillations lead to sequential steps in phosphorylation. Consequently, high frequency Ca^{2+} oscillations will fully activate the kinase quickly whereas slower Ca^{2+} oscillation will only partially activate the enzyme (Dupont and Goldbeter, 1998). Although FM regulation is a well accepted hypothesis, the number of clear examples describing this activity is surprisingly low.

6 Spatial Aspects of Ca^{2+} Signaling

An additional refinement to the fidelity of cell control exerted by Ca^{2+} is provided in the spatial domain of Ca^{2+} signaling. The Ca^{2+} signaling machinery of the cell is composed of multiple discrete units including the RyR, IP_3R, ER, PLC and SERCA Pumps. These elements frequently exist in clusters or are unevenly distributed throughout the cell and this leads to local or polarized Ca^{2+} signals.

6.1 Elemental Ca^{2+} Signals

The discrete nature of the RyRs and IP_3Rs is responsible for the elemental Ca^{2+} signaling referred to as Ca^{2+} sparks or Ca^{2+} puffs. These Ca^{2+} signals consist of brief, small and localized releases of Ca^{2+} from the ER. Ca^{2+} sparks are believed to arise from small clusters of RyRs whereas Ca^{2+} puffs arise at sites of IP_3Rs. The exact size and nature of the spark or puff will be influenced by the size of the cluster and it is not clear if the clusters of RyR or IP_3R are pure or can exist as mixtures. In addition, the rate at which the sparks or puffs occur is dependent on the cell or cytosolic conditions. Ca^{2+} sparks have been frequently observed in isolated cardiac cells and here they appear to result from the overloading of the SR. This result also highlights the influence of the Ca^{2+} concentration within the SR/ER on the open probability of the RyR. On the overhand, low frequency Ca^{2+} puffs are generally observed in response to low levels of cytoplasmic IP_3. It is not clear if the RyR agonist cADP-ribose can act in a similar way to stimulate sparks from the RyR.

6.2 Microdomains of Ca^{2+} Signaling

An extension of the elemental organization of the Ca^{2+} signaling is their relationship to each other. In the cardiac muscle, the RyRs are closely opposed to the L-type Ca^{2+} channels as this ensures high concentrations of Ca^{2+} are focused on the RyR while the global cytoplasm concentration changes little. Mitochondria are often involved in Ca^{2+} buffering and they are found in very close opposition to the ER, where they are able to perform Ca^{2+} uptake. Thus, by restricting the key Ca^{2+} signaling elements to small areas, signaling can be quickly accomplished without excessive increase in global Ca^{2+}.

6.3 Intracellular Ca^{2+} Waves

The elemental arrangements of RyRs or IP_3Rs throughout the cell also sets the foundation for the escalation of Ca^{2+} sparks or puffs into Ca^{2+} waves that travel across the entire cell (Coombes et al., 2004; Macrez and Mironneau, 2004) (Fig. 4). In this way, a signal detected by one part of the cell can be

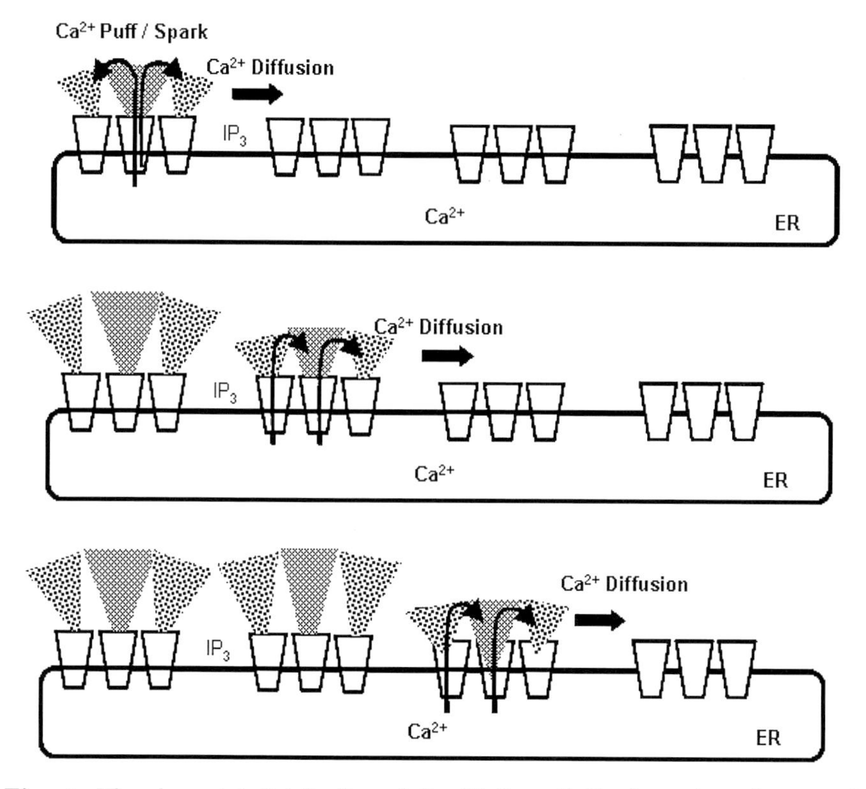

Fig. 4. The elemental distribution of the IP_3R or RyRs determines the spatial aspects of the Ca^{2+} signals in cells. Release of Ca^{2+} by a RyR cluster or an IP_3R cluster primed with IP_3 is followed by Ca^{2+} diffusion through the cytoplasm to an adjacent cluster. If the Ca^{2+} concentration is sufficient, further Ca^{2+} released is initiated and the process is repeated. CICR via RYR or IP_3-primed IP_3Rs initiates a step-like wave propagation that can extend from one end of the cell to the other

transmitted across the whole cell to initiate a homogenous response. The only requirement for wave propagation is the diffusion of Ca^{2+} between adjacent elemental release sites. Frequently, Ca^{2+} oscillations will be initiated in the same location within the cell and these spread as a wave to flood the cell with a tide of Ca^{2+}. The parameters that determine the characteristics of these waves are primarily the speed of Ca^{2+} diffusion and the number and distance between the elemental release sites. By simply reducing the spacing between sites, the wave propagation speed can be increased. The width of the Ca^{2+} wave-front is determined by inactivation kinetics of the release process and by the re-uptake of the Ca^{2+}. In most small cells, the time taken to inactivate release is slower than the time for the wave to propagate across the cell. The initiation and propagation of Ca^{2+} waves is also strongly influenced by the Ca^{2+} buffering capacity of the cytosol (Dargan et al., 2004). Fast buffers tend to reduce the likelihood that waves will propagate as this effective neutralizes the Ca^{2+} before it has had chance to diffuse. Fixed buffers may have a similar effect. The location and activity of the Ca^{2+} pumps will strongly influence the rate of the Ca^{2+} oscillations and the ability to propagate a Ca^{2+} wave.

6.4 Intercellular Ca^{2+} Waves

In higher organisms, cells are generally part of tissues and need to communicate with each other. In many cases, such communication can be between cells via gap junctions (Fig. 5). Gap junctions are constructed from 2 hemichannels that span the membrane of each cell to form a large conductance channel between the cells (Evans and Martin, 2002). Molecules of up to 1000 D in molecular weight can diffuse though a gap junction, although molecular charge and shape also contribute to the permeability properties. There are a wide variety of connexin isoforms and these are expressed by different cell types or by the same cells when in contact with different cells.

Gap junctions allow the passage of Ca^{2+} and IP_3. Consequently, Ca^{2+} signals can pass from one cell to the next. However, depending on the nature of the stimulus, the propagation mechanism of intercellular Ca^{2+} waves can vary. With local stimulation, where only one or a few cells display an elevation in IP_3, the communication of a Ca^{2+} wave to adjacent cells probably relies on the diffusion of IP_3 (Sanderson, 1995). In comparison to the diffusion of Ca^{2+}, the diffusion of IP_3 is almost unrestricted in the cytosol since there are very few IP_3 buffers. The major hindrance to IP_3 diffusion is IP_3 metabolism. The effectiveness of IP_3 diffusion is also enhanced by the fact that IP_3 initiates a saturating response in terms of Ca^{2+} release when it activates an IP_3R. Consequently, a uniform Ca^{2+} response can be initiated along the length of a declining diffusion gradient of IP_3 and, as a result, IP_3 diffusive waves can pass through multiple cells.

By contrast, local increases in Ca^{2+} spread poorly to adjacent cells and Ca^{2+} alone seems incapable of initiating a Ca^{2+} wave. The major reason for this, especially in the cell receiving a small influx of Ca^{2+} via a gap junction,

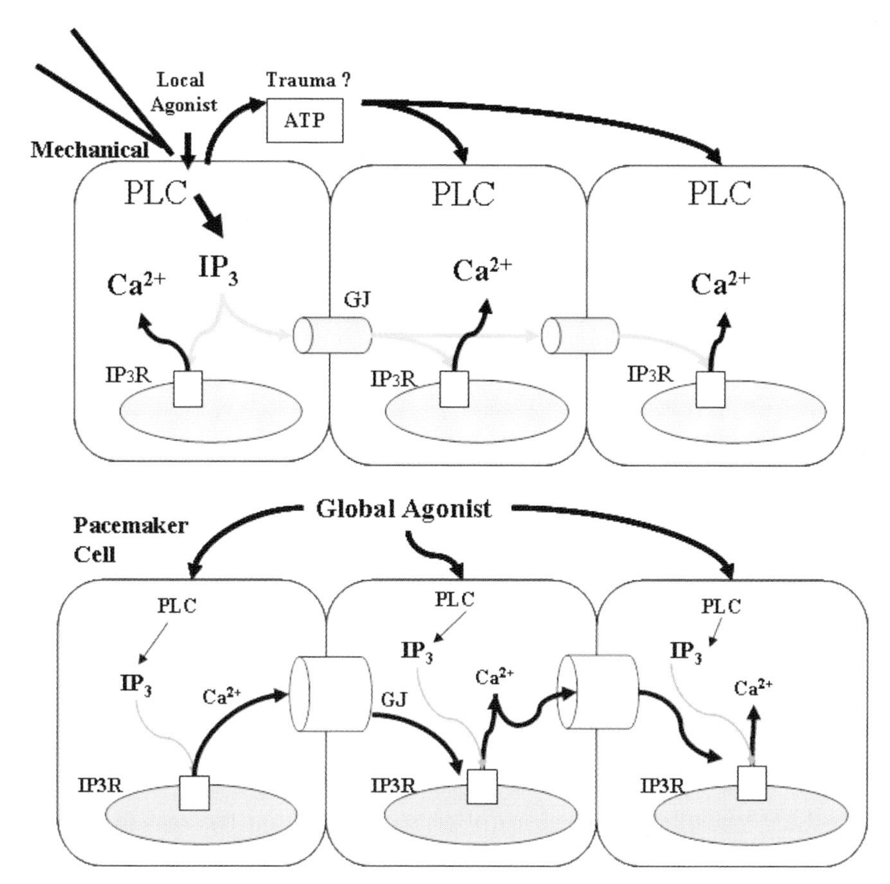

Fig. 5. Ca^{2+} waves can propagate between cells via gap junctions. (*Top*) A local agonist or mechanical stimulus results in the local production of IP_3. Diffusion of IP_3 through multiple cells stimulates the release of Ca^{2+} in each cell via the IP_3R. Because the source of the IP_3 is the stimulated cell, the wave is not regenerative. The release of an extracellular messenger, such as ATP, will generate a propagating wave as each cell is sequentially stimulated to produce IP_3. If ATP is released by each cell in response to the Ca^{2+} increase the wave will be regenerative. (*Bottom*) Ca^{2+} is a poor intercellular messenger and can only propagate a Ca^{2+} wave when the gap junction conductance between cells is good and the IP_3Rs of each cell have been primed with IP_3 following global stimulation with an agonist

is that Ca^{2+} buffers quickly prevent the free Ca^{2+} concentration from rising to stimulate RyRs of the next cell. However, under conditions of global hormonal stimulation, where all cells are experiencing some PLC activity and IP_3 production, the influx of Ca^{2+} via a gap junction, has a greater chance of stimulating Ca^{2+} release from a sensitized IP_3R to initiate a wave in the adjacent cell. The caveat to this mechanism is that the extent of the gap junctional coupling must be sufficiently high to allow a rate of Ca^{2+} flux that can stimulate the adjacent cell receptors. In many cell types, the extent of gap junctional coupling is low and this would support the observation that in response to agonist stimulation most cells only display asynchronous Ca^{2+} oscillations. Clearly, under these circumstances, the IP_3 receptors are primed and responding, but elevations in Ca^{2+} in one cell are not influencing adjacent cells.

6.5 Extracellular Ca^{2+} Waves

Although not strictly intercellular Ca^{2+} signaling, the diffusion of extracellular agonists can lead to propagating Ca^{2+} waves (Suadicani et al., 2004) (Fig. 5). In this respect, the release of ATP by cells has been of particular interest because ATP appears to leave the cell via uncoupled gap junction hemi-channels and this can easily be confused with intercellular Ca^{2+} waves passing through functional gap junctions. Because the Ca^{2+} signaling in each cell results from agonist activation of PLC and the production of IP_3 and that this IP_3 production stimulates the further release of ATP, these types of Ca^{2+} wave are regenerative and propagate over considerable numbers of cells. It is not unreasonable to expect that extracellular waves can propagate simultaneously with intercellular Ca^{2+} waves if the cells express connexins.

7 Summary

The Ca^{2+} signaling mechanisms of cells are based on the release of stored Ca^{2+}. Consequently, cells have developed a diversity of Ca^{2+} stores, Ca^{2+} pumps and Ca^{2+} release channels to control the movement of Ca^{2+}. By combining these elemental controls in different ways, different cells have the ability to generate signaling cascades that span a broad temporal or spatial range. However, this diversity rarely exists in an individual cell type and models of Ca^{2+} dynamics need to reflect the experimental details of a system to be of value.

References

Berridge MJ, Bootman MD, Roderick HL. (2003). Calcium signalling: dynamics, homeostasis and remodelling. *Nat Rev Mol Cell Biol* 4, 517–29.

Berridge MJ, Lipp P, Bootman MD. (2000). The versatility and universality of calcium signalling. *Nat Rev Mol Cell Biol* 1, 11–21.

Bootman MD, Berridge MJ, Roderick HL. (2002). Calcium signalling: more messengers, more channels, more complexity. *Curr Biol* 12, R563-5.

Coombes S, Hinch R, Timofeeva Y. (2004). Receptors, sparks and waves in a fire-diffuse-fire framework for calcium release. *Prog Biophys Mol Biol* 85, 197–216.

Dargan SL, Schwaller B, Parker I. (2004). Spatiotemporal patterning of IP3-mediated Ca2+ signals in Xenopus oocytes by Ca2+-binding proteins. *J Physiol* 556, 447–61.

Dupont G, Goldbeter A. (1998). CaM kinase II as frequency decoder of Ca2+ oscillations. *Bioessays* 20, 607–10.

Evans WH, Martin PE. (2002). Gap junctions: structure and function (Review). *Mol Membr Biol* 19, 121–36.

Hudmon A, Schulman H. (2002). Structure-function of the multifunctional Ca2+/calmodulin-dependent protein kinase II. *Biochem J* 364, 593–611.

Lee HC. (2004). Multiplicity of Ca2+ messengers and Ca2+ stores: a perspective from cyclic ADP-ribose and NAADP. *Curr Mol Med* 4, 227–37.

Macrez N, Mironneau J. (2004). Local Ca2+ signals in cellular signalling. *Curr Mol Med* 4, 263–75.

Meissner G. (2004). Molecular regulation of cardiac ryanodine receptor ion channel. *Cell Calcium* 35, 621–8.

Nilius B, Voets T. (2004). Diversity of TRP channel activation. *Novartis Found Symp* 258, 140–9; discussion 149–59, 263–6.

Petersen OH, Tepikin A, Park MK. (2001). The endoplasmic reticulum: one continuous or several separate Ca(2+) stores? *Trends Neurosci* 24, 271–6.

Putney JW, Jr. (2004). Store-operated calcium channels: how do we measure them, and why do we care? *Sci STKE* 2004, 37.

Sanderson MJ. (1995). Intercellular calcium waves mediated by inositol trisphosphate. *Ciba Found Symp* 188, 175–89; discussion 189–94.

Shuttleworth TJ. (2004). Receptor-activated calcium entry channels–who does what, and when? *Sci STKE* 2004, 40.

Strehler EE, Treiman M. (2004). Calcium pumps of plasma membrane and cell interior. *Curr Mol Med* 4, 323–35.

Suadicani SO, Flores CE, Urban-Maldonado M, Beelitz M, Scemes E. (2004). Gap junction channels coordinate the propagation of intercellular Ca(2+) signals generated by P2Y receptor activation. *Glia* 48, 217.

Swann K, Larman MG, Saunders CM, Lai FA. (2004). The cytosolic sperm factor that triggers Ca2+ oscillations and egg activation in mammals is a novel phospholipase C: PLCzeta. *Reproduction* 127, 431–9.

Taylor CW, da Fonseca PC, Morris EP. (2004). IP(3) receptors: the search for structure. *Trends Biochem Sci* 29, 210–9.

Modeling IP_3-Dependent Calcium Dynamics in Non-Excitable Cells

J. Sneyd

University of Auckland, New Zealand
sneyd@math.auckland.ac.nz

1 Introduction

Calcium is critically important for a vast array of cellular functions, as discussed in detail in Chap. 1. There are a number of Ca^{2+} control mechanisms operating on different levels, all designed to ensure that Ca^{2+} is present in sufficient quantity to perform its necessary functions, but not in too great a quantity in the wrong places. Prolonged high concentrations of Ca^{2+} are toxic. For instance, it is known that cellular Ca^{2+} overload can trigger apoptotic cell death, a process in which the cell kills itself. Indeed, control of Ca^{2+} homeostasis is so crucial that even just disruptions in the normal Ca^{2+} fluxes can lead to initiation of active cell death. There are many reviews of Ca^{2+} physiology in the literature: in 2003 an entire issue of Nature Reviews was devoted to the subject and contains reviews of Ca^{2+} homeostasis (Berridge et al., 2003), extracellular Ca^{2+} sensing (Hofer and Brown, 2003), Ca^{2+} signaling during embryogenesis (Webb and Miller, 2003), the Ca^{2+}-apoptosis link (Orrenius et al., 2003), and the regulation of cardiac contractility by Ca^{2+} (MacLennan and Kranias, 2003). Other useful reviews are Berridge, (1997) and Carafoli (2002).

In vertebrates, the majority of body Ca^{2+} is stored in the bones, whence it can be released by hormonal stimulation to maintain an extracellular Ca^{2+} concentration of around 1 mM, while the intracellular concentration is kept at around 0.1 μM. Since the internal concentration is low, there is a steep concentration gradient from the outside of a cell to the inside. This disparity has the advantage that cells are able to raise their Ca^{2+} concentration quickly, by opening Ca^{2+} channels and relying on passive flow down a steep concentration gradient, but it has the disadvantage that energy must be expended to keep the cytosolic Ca^{2+} concentration low. Thus, cells have finely tuned mechanisms to control the influx and removal of cytosolic Ca^{2+}.

Calcium is removed from the cytoplasm in two principal ways: it is pumped out of a cell, and it is sequestered into internal membrane-bound compartments such as the mitochondria, the endoplasmic reticulum (ER) or sarcoplasmic reticulum (SR), and secretory granules. Since the Ca^{2+} concentration in the cytoplasm is much lower than either the extracellular concentration or the concentration inside the internal compartments, both methods of Ca^{2+} removal require expenditure of energy. Some of this is by a Ca^{2+} ATPase, similar to the Na^+–K^+ ATPase, that uses energy stored in ATP to pump Ca^{2+} out of the cell or into an internal compartment. There is also a Na^+–Ca^{2+} exchanger in the cell membrane that uses the energy of the Na^+ electrochemical gradient to remove Ca^{2+} from the cell at the expense of Na^+ entry.

Calcium influx also occurs via two principal pathways: inflow from the extracellular medium through Ca^{2+} channels in the surface membrane and release from internal stores. The surface membrane Ca^{2+} channels are of several different types: voltage-controlled channels that open in response to depolarization of the cell membrane, receptor-operated channels that open in response to the binding of an external ligand, second-messenger-operated channels that open in response to the binding of a cellular second messenger, and mechanically operated channels that open in response to mechanical stimulation. Voltage-controlled Ca^{2+} channels are of great importance in muscle and are discussed in detail in Chap. 4. We also omit the consideration of the other surface membrane channels to concentrate on the properties of Ca^{2+} release from internal stores.

Calcium release from internal stores such as the ER is the second major Ca^{2+} influx pathway, and this is mediated principally by two types of Ca^{2+} channels that are also receptors: the ryanodine receptor and the inositol (1,4,5)-trisphosphate (IP_3) receptor. The ryanodine receptor, so-called because of its sensitivity to the plant alkaloid ryanodine, plays an integral role in excitation–contraction coupling in skeletal and cardiac muscle cells, and is believed to underlie Ca^{2+}-induced Ca^{2+} release, whereby a small amount of Ca^{2+} entering the cardiac cell through voltage-gated Ca^{2+} channels initiates an explosive release of Ca^{2+} from the sarcoplasmic reticulum. We discuss excitation-contraction coupling in detail in Chap. 12. Ryanodine receptors are also found in a variety of nonmuscle cells such as neurons, pituitary cells, and sea urchin eggs. The IP_3 receptor, although similar in structure to the ryanodine receptor, is found predominantly in nonmuscle cells, and is sensitive to the second messenger IP_3. The binding of an extracellular agonist such as a hormone or a neurotransmitter to a receptor in the surface membrane can cause, via a G-protein link to phospholipase C (PLC), the cleavage of phosphotidylinositol (4,5)-bisphosphate (PIP_2) into diacylglycerol (DAG) and IP_3 (Fig. 1). The water-soluble IP_3 is free to diffuse through the cell cytoplasm and bind to IP_3 receptors situated on the ER membrane, leading to the opening of these receptors and subsequent release of Ca^{2+} from the ER. Similarly to ryanodine receptors, IP_3 receptors are modulated by the cytosolic Ca^{2+} concentration, with Ca^{2+} both activating and inactivating Ca^{2+} release, but at different rates. Thus Ca^{2+}-induced Ca^{2+} release occurs through IP_3 receptors also.

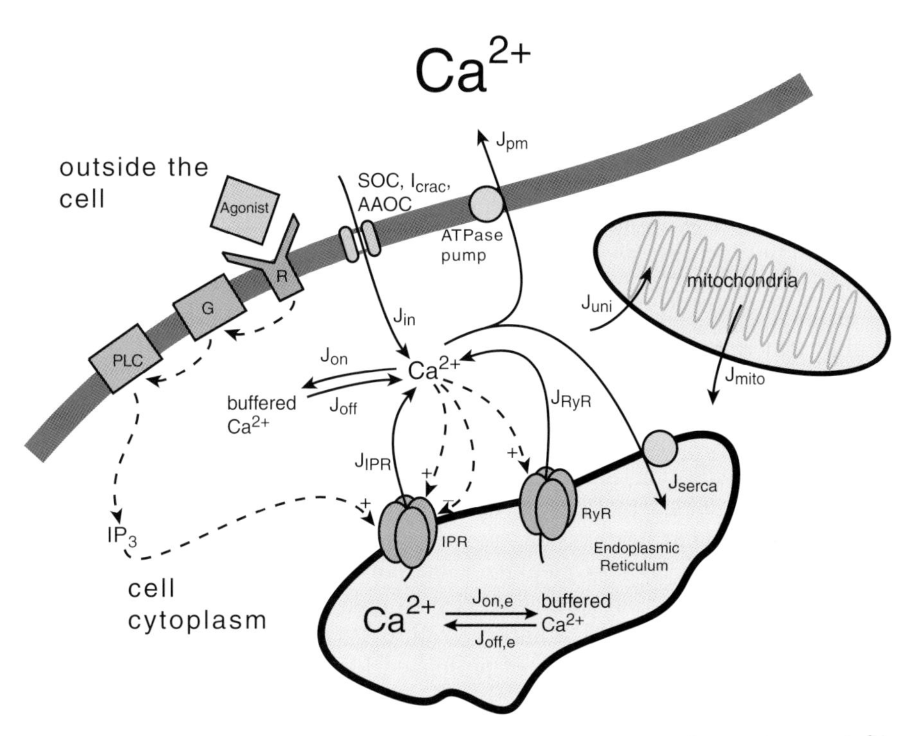

Fig. 1. Diagram of the major fluxes involved in the control of cytoplasmic Ca^{2+} concentration. Binding of agonist to a cell membrane receptor (R) leads to the activation of a G-protein (G), and subsequent activation of phospholipase C (PLC). This cleaves phosphotidylinositol bisphosphate (PIP_2) into diacylglycerol and inositol trisphosphate (IP_3), which is free to diffuse through the cell cytoplasm. When IP_3 binds to an IP_3 receptor (IPR) on the endoplasmic reticulum (ER) membrane it causes the release of Ca^{2+} from the ER, and this Ca^{2+} in turn modulates the open probability of the IPR and ryanodine receptors (RyR). Calcium fluxes are denoted by *solid* arrows. Calcium can be released from the ER through IPR (J_{IPR}) or RyR (J_{RyR}), can be pumped from the cytoplasm into the ER (J_{serca}) or to the outside (J_{pm}), can be taken up into (J_{uni}), or released from (J_{mito}), the mitochondria, and can be bound to (J_{on}), or released from (J_{off}), Ca^{2+} buffers. Entry from the outside is controlled by a variety of possible channels, included store-operated channels (SOC), calcium release activated channels (I_{crac}) and arachidonic acid operated channels (AAOC)

As an additional control for the cytosolic Ca^{2+} concentration, Ca^{2+} is heavily buffered (i.e., bound) by large proteins, with estimates that at least 99% of the total cytoplasmic Ca^{2+} is bound to buffers. The Ca^{2+} in the ER and mitochondria is also heavily buffered.

2 Calcium Oscillations and Waves

One of the principal reasons that modelers have become interested in calcium dynamics is that the concentration of Ca^{2+} shows highly complex spatio-temporal behaviour. In response to agonists such as hormones or neurotransmitters, many cell types exhibit oscillations in intracellular $[Ca^{2+}]$. These oscillations can be grouped into two major types: those that are dependent on periodic fluctuations of the cell membrane potential and the associated periodic entry of Ca^{2+} through voltage-gated Ca^{2+} channels, and those that occur in the presence of a voltage clamp. Our focus here is on the latter type, within which group further distinctions can be made by whether the oscillatory calcium flux is through ryanodine or IP_3 receptors. We consider only models of the IP_3 receptors here. Ryanodine receptor models are discussed in more detail in Chaps. 3 and 4.

Calcium oscillations have been implicated in a vast array of cellular control processes. The review by Berridge et al. (2003) mentions, among other things, oocyte activation at fertilization, axonal growth, cell migration, gene expression, formation of nodules in plant root hairs, development of muscle, and release of cytokines from epithelial cells. In many cases the oscillations are a frequency-encoded signal that allows a cell to use Ca^{2+} as a second messenger while avoiding the toxic effects of prolonged high $[Ca^{2+}]$. For example, exocytosis in gonadotropes is known to be dependent on the frequency of Ca^{2+} oscillations (Tse et al., 1993), and gene expression can also be modulated by Ca^{2+} spike frequency (Dolmetsch et al., 1998; Li et al., 1998). However, there are still many examples where the signal carried by a Ca^{2+} oscillation has not yet been unambiguously decoded.

The period of IP_3-dependent oscillations ranges from a few seconds to a few minutes (Fig. 2). In some cell types these oscillations appear to occur at a constant concentration of IP_3, while in other cell types they appear to be driven by oscillations in $[IP_3]$. Modulation of the IP_3 receptor by other factors such as phosphatases and kinases also plays an important role in setting the oscillation period, while control of Ca^{2+} influx from outside the cell appears to be an important regulatory mechanism. There is thus a tremendous variety of mechanisms that control Ca^{2+} oscillations, and it is not realistic to expect that a single model will capture the important behaviours in all cell types. Nevertheless, most of the models have much in common, and a great deal can be learned about the overall approach by the study of a small number of models. The concept of a calcium-signaling "toolkit" has been introduced by Berridge et al. (2003). There are a large number of components in the toolkit (receptors, G proteins, channels, buffers, pumps, exchangers and so on); by expressing just those components which are needed, each cell can fine-tune the spatio-temporal properties of intracellular Ca^{2+}.

Often Ca^{2+} oscillations do not occur uniformly throughout the cell, but are organized into repetitive intracellular waves (Rooney and Thomas, 1993; Thomas et al., 1996; Røttingen and Iversen, 2000; Falcke 2004). One of the

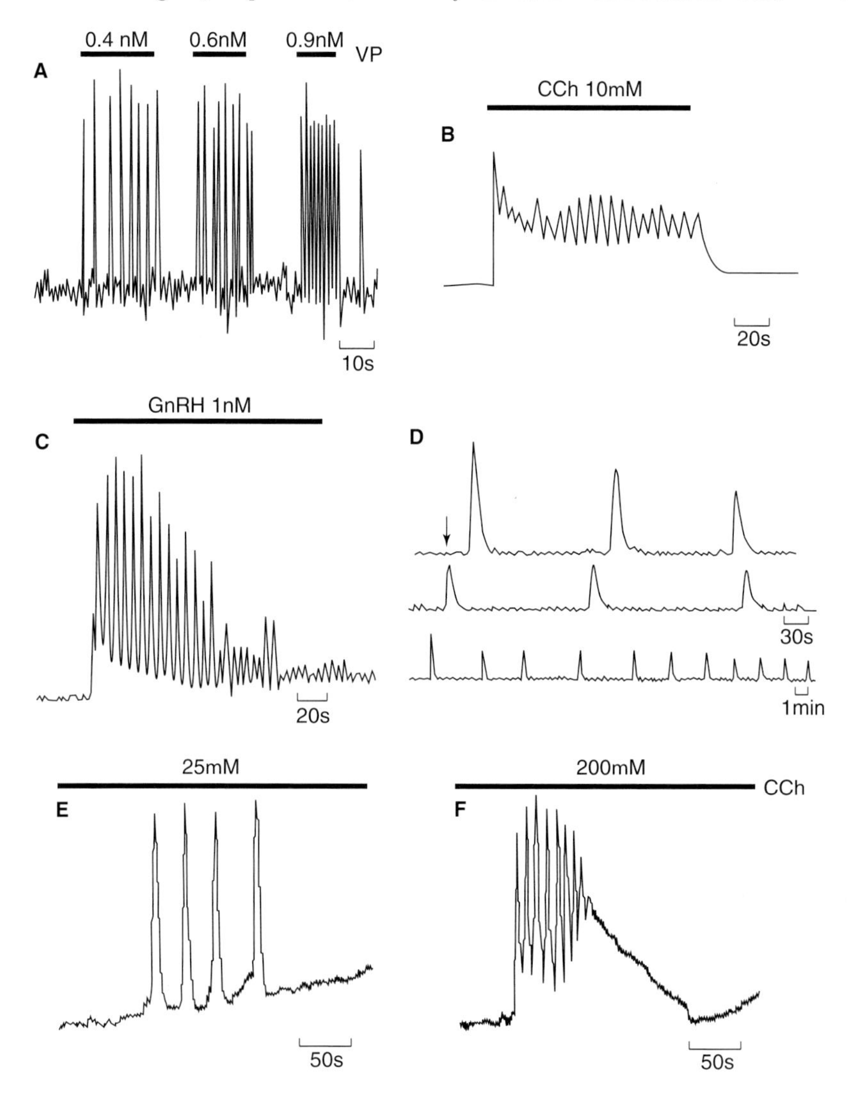

Fig. 2. Typical calcium oscillations from a variety of cell types. (**A**) Hepatocytes stimulated with vasopressin (VP). (**B**) Rat parotid gland stimulated with carbachol (CCh). (**C**) Gonadotropes stimulated with gonadotropin-releasing hormone (GnRH). (**D**) Hamster eggs after fertilization. The time of fertilization is denoted by the *arrow*. (**E**) and (**F**) Insulinoma cells stimulated with two different concentrations of carbachol (Berridge and Galione, 1998, Fig. 2.)

most visually impressive examples of this occurs in *Xenopus* oocytes. In 1991, it was discovered by Lechleiter and Clapham and their coworkers that intracellular Ca^{2+} waves in immature *Xenopus* oocytes showed remarkable spatiotemporal organization. By loading the oocytes with a Ca^{2+}-sensitive dye, releasing IP_3, and observing Ca^{2+} release patterns with a confocal microscope, (Lechleiter and Clapham, 1992; Lechleiter et al., 1991a,b) observed that the intracellular waves develop a high degree of spatial organization, forming concentric circles, plane waves, and multiple spiral Ca^{2+} waves in vivo. Typical experimental results are shown in Fig. 3A. The crucial feature of *Xenopus* oocytes that makes these observations possible is their large size. *Xenopus* oocytes can have a diameter larger than 600 µm, an order of magnitude greater than most other cells. In a small cell, a typical Ca^{2+} wave (often with a width of close to 100 µm) cannot be observed in its entirety, and there is not enough room for a spiral to form. However, in a large cell it may be possible to observe both the wave front and the wave back, as well as spiral waves, and this has made the *Xenopus* oocyte one of the most important systems for the study of Ca^{2+} waves. Of course, what is true for *Xenopus* oocytes is not necessarily true for other cells, and so one must be cautious about extrapolating to other cell types.

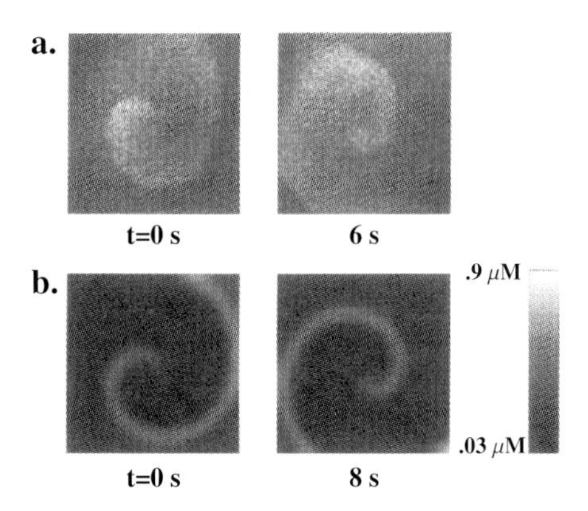

Fig. 3. (**A**) Spiral Ca^{2+} wave in the *Xenopus* oocyte. The image size is 420×420 µm. The spiral has a wavelength of about 150 µm and a period of about 8 seconds. (**B**) A model spiral wave simulated on a domain of size 250 × 250 µm, with $[IP_3] = 95$ nM. See Sect. 4.1 (Atri et al., 1993, Fig. 11.)

Another well-known example of an intracellular Ca^{2+} wave occurs across the cortex of an egg directly after fertilization (Ridgway et al., 1997; Nuccitelli et al., 1993; Fontanilla and Nuccitelli, 1998). In fact, this wave motivated the very first models of Ca^{2+} wave propagation, which appeared as early as 1978 (Gilkey et al., 1978; Cheer et al., 1997; Lane et al., 1987). These early models, unsurprisingly, have since been superseded (Wagner et al., 1998; Bugrim et al., 2003).

In addition to traveling across single cells, Ca^{2+} waves can be transmitted between cells, forming intercellular waves that can travel over distances of many cell lengths. Such intercellular waves have been observed in intact livers, slices of hippocampal brain tissue, epithelial and glial cell cultures (see Fig. 4), and many other preparations (Sanderson et al., 1990; Sanderson et al., 1994; Charles et al., 1991, 1992; Robb-Gaspers and Thomas, 1995). Not all intercellular coordination is of such long range; synchronised oscillations are often observed in small groups of cells such as pancreatic or parotid acinar cells (Yule et al., 1996) or multiplets of hepatocytes (Tordjmann et al., 1997, 1998).

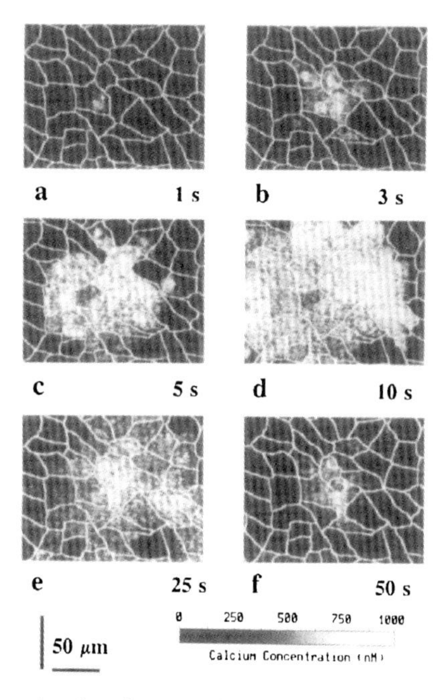

Fig. 4. Mechanically stimulated intercellular wave in airway epithelial cells. The time after mechanical stimulation is given in seconds in the *lower right* corner of each panel (Sneyd et al., 1995a, Fig. 4A.)

Although there is controversy about the exact mechanisms by which Ca^{2+} waves propagate (and it is certainly true that the mechanisms differ from cell type to cell type), it is widely believed that in many cell types, intracellular Ca^{2+} waves are driven by the diffusion of Ca^{2+} between Ca^{2+} release sites. According to this hypothesis, the Ca^{2+} released from one group of release sites (usually either IPR or RyR) diffuses to neighboring release sites and initiates further Ca^{2+} release from them. Repetition of this process can generate an advancing wave front of high Ca^{2+} concentration, i.e., a Ca^{2+} wave. Since they rely on the active release of Ca^{2+} via some positive feedback mechanism, such waves are called actively propagated. The theory of such waves is presented in detail in Keener and Sneyd (1998). However, when the underlying Ca^{2+} kinetics are oscillatory (i.e., there is a stable limit cycle), waves can propagate by a kinematic, or phase wave, mechanism, in which the wave results from the spatially ordered firing of local oscillators. Such waves do not depend on Ca^{2+} diffusion for their existence, merely on the fact that one end of the cell is oscillating with a different phase than the other end. In this case, Ca^{2+} diffusion serves to phase lock the local oscillators, but the phase wave can persist in the absence of Ca^{2+} diffusion.

The most common approach to the study of Ca^{2+} oscillations and waves is to assume that the underlying mechanisms are deterministic. However, even the most cursory examination of experimental data shows that Ca^{2+} dynamics is inherently stochastic in nature. The most prominent of the stochastic events underlying Ca^{2+} oscillations and waves are small localized release events called *puffs*, or *sparks*, and these elementary events, caused by the opening of single, or a small number of, Ca^{2+} release channels, are the building blocks from which global events are built. Calcium puffs, caused by localized release through IPR, have been extensively studied in *Xenopus* oocytes and HeLa cells (Marchant et al., 1999; Sun et al., 1998; Callamaras et al., 1998; Marchant and Parker, 2001; Thomas et al., 2000; Bootman et al., 1997a,b), while calcium sparks, caused by localized release through RyR and occurring principally in muscle cells, were discovered by Cheng et al. (1993, 1996) and studied by a multitude of authors since (Smith et al., 1998; Izu et al., 2001; Sobie et al., 2002; Soeller and Cannell, 1997, 2002). To study the stochastic properties of puffs and sparks, stochastic models are necessary. In general such models are based on stochastic simulations of Markov state models for the IPR and RyR. We discuss such models briefly in Sect. 6.

A wide variety of different models have been constructed to study Ca^{2+} oscillations and waves in many different cell types. There is not space enough here to discuss all the models in the literature. Thus, after discussing particular models for each of the most important Ca^{2+} fluxes, we shall discuss in detail only a few models for oscillations and waves. The most comprehensive review of the field is Falcke (2004) and the interested reader is referred to that review (which is almost 200 pages long and certainly the best review yet written of models of Ca^{2+} dynamics) for additional information. Another excellent review of models is that of Schuster et al. (2002).

3 Well-Mixed Cell Models: Calcium Oscillations

If we assume the cell is well-mixed, then the concentration of each species is homogeneous throughout. We write c for the concentration of free calcium ions in the cytoplasm and note that $c = c(t)$, i.e., c has no spatial dependence. Similarly, we let c_e denote the concentration of Ca^{2+} in the ER. The differential equations for c and c_e come from consideration of the fluxes in Fig. 1.

The differential equations for c and c_e follow simply from conservation of calcium. Thus, from consideration of the fluxes shown in Fig. 1 we have

$$\frac{dc}{dt} = J_{\mathrm{IPR}} + J_{\mathrm{RyR}} + J_{\mathrm{in}} - J_{\mathrm{pm}} - J_{\mathrm{serca}} - J_{\mathrm{on}} + J_{\mathrm{off}} + J_{\mathrm{uni}} - J_{\mathrm{mito}} \quad (1)$$

$$\frac{dc_e}{dt} = \gamma(J_{\mathrm{serca}} - J_{\mathrm{IPR}} - J_{\mathrm{RyR}}) + J_{\mathrm{off,e}} - J_{\mathrm{on,e}} \ , \quad (2)$$

where γ is the ratio of the cytoplasmic volume to the ER volume. The factor γ is necessary because c_e is in units of micromoles per liter ER, while c is in units of micromoles per liter cytoplasm. Since each flux to or from the cytoplasm is assumed to have units of micromoles per liter cytosol per second, the fluxes in and out of the ER must be scaled to take the different volumes into account. Each of the fluxes in this equation corresponds to a component of the Ca^{2+}-signaling toolkit; clearly, depending on what assumptions are made, the list of fluxes could be extended practically indefinitely. Here, we discuss only those that most often appear in models.

In general, these equations would be coupled to differential equations that model the IPR or RyR, or to additional equations describing the dynamics of the pumps and exchangers. An example of such a model is discussed in Sect. 3.6. However, before we see what happens when it all gets put together, let us first discuss how each toolkit component is modelled.

3.1 Influx

In general, the influx of Ca^{2+} into the cell from outside is voltage-dependent. However, when Ca^{2+} oscillations occur at a constant voltage (as is the norm in non-excitable cells) the voltage dependence is unimportant. This influx is certainly dependent on a host of factors, including Ca^{2+}, IP_3, arachidonic acid and the Ca^{2+} concentration in the ER. Very little is understood about how exactly this influx is controlled, so any detailed model of the influx would be highly ambitious at this stage. For instance, there is evidence that, in some cell types, depletion of the ER causes an increase in Ca^{2+} influx through store-operated channels, or SOCs. There is also evidence that SOCs play a role only at high agonist concentration (when the ER is highly depleted) but that at lower agonist concentrations Ca^{2+} influx is controlled by arachidonic acid (Shuttleworth, 1999). However, the exact mechanism by which this occurs is unknown. What we know is that J_{in} increases as agonist concentration increases; if this were not so, the steady-state level of Ca^{2+} would be independent of IP_3 which we know not to be the case. One common model thus makes the simple assumption that J_{leak} is a linear increasing function of p, the IP_3 concentration. Thus

$$J_{\mathrm{in}} = \alpha_1 + \alpha_2 p \ , \quad (3)$$

for some constants α_1 and α_2.

3.2 Mitochondria

Mitochondrial Ca^{2+} handling is a highly complex field, and a number of detailed models have been constructed. For the sake of brevity we shall not discuss these models at all; the interested reader is referred to Colegrove et al. (2000), Friel (2000), Falcke et al. (2000), Grubelnik et al. (2001), Marhl et al. (2000), Schuster et al. (2002) and Selivanov et al. (1998). It seems that one function of the mitochondria is to take up and release large amounts of Ca^{2+}, but relatively slowly. Thus, the mitochondria tend to modulate the trailing edges of the waves, reduce wave amplitude, and change the long-term oscillatory behaviour. However, this is certainly an over-simplification.

3.3 Calcium Buffers

At least 99% of Ca^{2+} in the cytoplasm is bound to large proteins, called Ca^{2+} buffers. Typical buffers are calsequestrin, calbindin, flourescent dyes, and the plasma membrane itself. A detailed discussion of Ca^{2+} buffering, and its effect on oscillations and waves, is given in Sect. 5.

3.4 Calcium Pumps and Exchangers

3.4.1 Calcium ATPases

Early models of the calcium ATPase pump were based simply on the Hill equation (cf. Keener and Sneyd, 1998). For instance, data from Lytton et al. (1992) showed that the flux through the ATPase was approximately a sigmoidal function of c, with Hill coefficient of about 2. Thus, a common model is simply to set

$$J_{\text{serca}} = \frac{V_p c^2}{K_p^2 + c^2} \,. \tag{4}$$

However, such simple models have a number of serious flaws. Firstly, they assume that the flux is only ever in one direction, which we know is not the case; given a high enough ER calcium concentration, it is possible for the pump to reverse, generating ATP in the process. Secondly, it is not obvious how ER calcium concentrations can affect the flux, although we know that such effects are important.

MacLennan et al. (1997) discuss a more accurate model, shown schematically in Fig. 5. The pump can be in one of two basic conformations: E_1 and E_2. In the E_1 conformation the pump can bind two Ca^{2+} ions from the cytoplasm, whereupon it exposes a phosphorylation site. Once phosphorylated, the pump can switch to the E_2 conformation in which the Ca^{2+} binding sites are exposed to the ER lumen and have a much lower affinity. Thus Ca^{2+} is released into the ER, the pump is dephosphorylated, and finally completes the cycle by switching back to the E_1 conformation. For each Ca^{2+} ion transported from the cytoplasm to the ER, one proton is co-transported from the

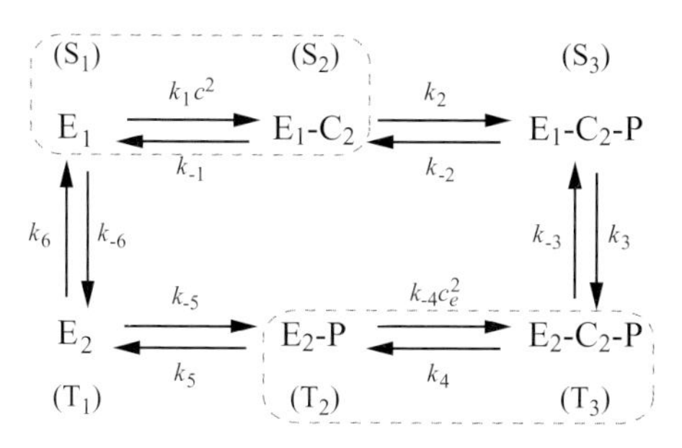

Fig. 5. Schematic diagram of the SERCA model of MacLennan et al. (1997). E_1 is the conformation where the Ca^{2+} binding sites are exposed to the cytoplasm, and E_2 is the conformation where they are exposed to the lumen and have a much lower affinity. The P denotes that the pump has been phosphorylated; for simplicity we have omitted ATP and ADP from the diagram. We have also omitted the co-transport of H^+. By assuming fast equilibrium between the pairs of states in the *dashed* boxes we can derive a simplified model, as shown in Fig. 6

ER to the cytoplasm. This makes the similarity with the Na^+/K^+ ATPase even more apparent. (In the models that we discuss here we shall ignore any effects of H^+ concentration on the pump rate. However, it is important to keep in mind that such proton transport will be an important feature to consider in any model of cellular pH control.) The rate limiting step of the transport cycle is the transition from E_1-C_2-P to E_2-C_2-P.

To calculate the flux in such a model is relatively straightforward, although the algebra can get messy and is best done on a computer. For convenience we label the pump states S_i and T_i, $i = 1, 2, 3$, as shown in Fig. 5, and let $s_i(t_i)$ denote the fraction of pumps in state $S_i(T_i)$. Because of the conservation constraint

$$\sum_{i=1}^{3}(s_i + t_i) = 1 \tag{5}$$

only five of these variables are independent.

At steady state, the flux of Ca^{2+} from the cytoplasm to the ER is given by the difference between the forward and reverse rates of any of the reaction steps. (Note that this difference must be same for each reaction step otherwise the model would not be at steady state.) Thus, if we let J denote the Ca^{2+} flux from the cytoplasm to the ER, we have

$$\begin{aligned}
J &= k_1 s_1 c^2 - k_{-1} s_2 && (6)\\
&= k_2 s_2 - k_{-2} s_3 && (7)\\
&= k_3 s_3 - k_{-3} t_3 && (8)\\
&= k_4 t_3 - k_{-4} t_2 c_e^2 && (9)\\
&= k_5 t_2 - k_{-5} t_1 && (10)\\
&= k_6 t_1 - k_{-6} s_1 \ , && (11)
\end{aligned}$$

where c denotes $[\mathrm{Ca}^{2+}]$ in the cytoplasm and c_e denotes $[\mathrm{Ca}^{2+}]$ in the ER. We thus have six equations for six unknowns (s_1, s_2, s_3, t_1, t_2 and J), which can be solved to find J as a function of the rate constants.

Solution of these equations gives

$$J = \frac{c^2 - K_1K_2K_3K_4K_5K_6c_e^2}{\alpha_1c^2 + \alpha_2c_e^2 + \alpha_3c^2c_e^2 + \alpha_4}, \tag{12}$$

where the αs are functions of the rate constants, too long and of too little interest to include in full detail. As always, $K_i = k_{-i}/k_i$. If the affinity of the Ca^{2+} binding sites is high when the pump is in conformation E_1, it follows that K_2 is small (i.e., $k_{-2} \ll k_2$). Similarly, if the affinity of the binding sites is low when the pump is in conformation E_2, it follows that K_4 is also small. (Note that, given the way we have written the diagram, k_{-4} is the rate at which ER Ca^{2+} binds to the pump in conformation E_2.) Since K_2 and K_4 are thus small, it follows that the pump can support a positive flux even when c_e is much greater than c.

If this were a model of a closed system, the law of detailed balance would require that $K_1K_2K_3K_4K_5K_6 = 1$. Fortunately, this is not so here. Because the cycle is driven by phosphorylation of the pump, the reaction rates depend upon the concentrations of ATP, ADP and P, and energy is continually being consumed (if $J > 0$) or generated (if $J < 0$).

In the original description of the model, MacLennan et al. assumed that the binding and release of Ca^{2+} occurred quickly. If we make this assumption we get a similar expression for the steady-state flux. However, the process of how one reduces the model in this way is important and useful, and thus worth considering in detail.

In Fig. 5 states S_1 and S_2 have been grouped together by a box with a dotted outline, and so have states T_2 and T_3. If we assume that these pairs of states are in instantaneous equilibrium it follows that

$$c^2s_1 = K_1s_2 \tag{13}$$

$$t_3 = K_4c_e^2t_2 . \tag{14}$$

Now let

$$\bar{s}_1 = s_1 + s_2 \tag{15}$$

$$= s_1\left(1 + \frac{c^2}{K_1}\right) \tag{16}$$

$$= s_2\left(1 + \frac{K_1}{c^2}\right), \tag{17}$$

and

$$\bar{t}_2 = t_2 + t_3 \tag{18}$$

$$= t_2(1 + K_4c_e^2) \tag{19}$$

$$= t_3\left(1 + \frac{1}{K_4c_e^2}\right). \tag{20}$$

Thus

$$\frac{d\bar{s}_1}{dt} = k_{-2}s_3 - k_2 s_2 + k_6 t_1 - k_{-6}s_1 \tag{21}$$

$$= k_{-2}s_3 - \left(\frac{k_2}{1+\frac{K_1}{c^2}}\right)\bar{s}_1 + k_6 t_1 - \left(\frac{k_{-6}}{1+\frac{c^2}{K_1}}\right)\bar{s}_1 \tag{22}$$

$$\frac{d\bar{t}_2}{dt} = k_3 s_3 - k_{-3}t_3 + k_{-5}t_1 - k_5 t_2 \tag{23}$$

$$= k_3 s_3 - \left(\frac{k_{-3}}{1+\frac{1}{K_4 c_e^2}}\right)\bar{t}_2 + k_{-5}t_1 - \left(\frac{k_5}{1+K_4 c_e^2}\right)\bar{t}_2 \, . \tag{24}$$

From these equations we can write down a simplified schematic diagram of the ATPase pump, as shown in Fig. 6. As c increases, the equilibrium between S_1 and S_2 will be shifted further towards S_2. This will tend to increase the rate at which S_3 is formed, but decrease the rate at which S_1 is converted to T_1. Thus, the transition from $\bar{S}_1$ to S_3 is an increasing function of c, while the transition from $\bar{S}_1$ to T_1 is a decreasing function of c.

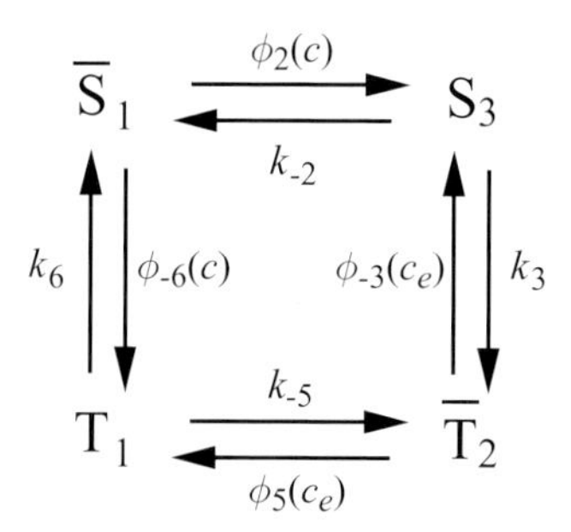

Fig. 6. Schematic diagram of the simiplified version of the SERCA model of MacLennan et al. (1997). By assuming fast equilibrium between the pairs of states shown grouped in Fig. 5 by the *dashed* boxes, we can derive this diagram in which there are fewer states, but the transitions are nonlinear functions of c. The functions in the transition rates are $\phi_2 = c^2 k_2/(c^2 + K_1)$, $\phi_{-3} = k_{-3}K_4 c_e^2/(1 + K_4 c_e^2)$, $\phi_5 = k_5/(1 + K_4 c_e^2)$ and $\phi_{-6} = K_1 k_{-6}/(K_1 + c^2)$

From this diagram we can calculate the steady-state flux in the same way as before. Not surprisingly, the final result looks much the same, giving

$$J = \frac{c^2 - K_1K_2K_3K_4K_5K_6c_e^2}{\beta_1 c^2 + \beta_2 c_e^2 + \beta_3 c^2 c_e^2 + \beta_4} \, , \tag{25}$$

for some constants β that are not the same as the constants α that appeared previously.

The calcium ATPase pumps on the plasma membrane are similar to the SERCA ATPase pumps, and are modeled in the same way.

3.4.2 Calcium Exchangers

Another important way in which Ca^{2+} is removed from the cytoplasm is via the action of Na^+-Ca^{2+} exchangers, which remove one Ca^{2+} ion from the cytoplasm at the expense of the entry of three Na^+ ions. They are particularly important for the control of Ca^{2+} in muscle and are discussed in more detail in Chaps. 3 and 4.

3.5 IP_3 Receptors

The basic property of IP_3 receptors is that they respond in a time-dependent manner to steps of Ca^{2+} or IP_3. Thus, in response to a step increase of IP_3 or Ca^{2+} the receptor open probability first opens to a peak and then declines to a lower plateau (see Fig. 7). This decline is called *adaptation* of the receptor, as it adapts to a maintained Ca^{2+} or IP_3 concentration. If a further step is applied on top of the first one, the receptor will respond with another peak, followed

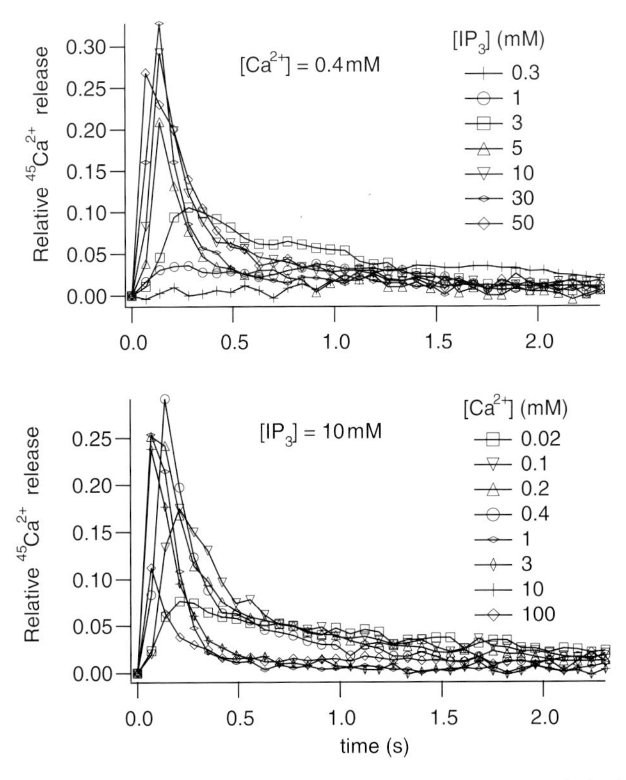

Fig. 7. Experimental data from Dufour et al., 1997. Response of the IPR to step increases in IP_3 or Ca^{2+}

by a decline to a plateau. In this way the IPR responds to *changes* in Ca^{2+} or IP_3 concentration, rather than to the absolute background concentration.

Adaptation of the IPR is now believed to result, at least in part, from the fact that, not only does Ca^{2+} stimulate its own release, it also inhibits it, but on a slower time scale (Parker and Ivorra 1990; Finch et al., 1991; Bezprozvanny et al., 1991; Parys et al., 1992). It is hypothesized that this sequential activation and inactivation of the IP_3 receptor by Ca^{2+} is one mechanism underlying IP_3-dependent Ca^{2+} oscillations and waves, and a number of models incorporating this hypothesis have appeared (reviewed by Sneyd et al. (1995b); Tang et al., 1996; Schuster et al. (2002); Falcke (2004)). However, as we shall see, there are almost certainly other important mechanisms operating; some of the most recent models (Sneyd et al., 2004a; Sneyd and Falcke, 2004) have suggested that the IPR kinetics are of much less importance than previously thought, while depletion of the ER might also be playing in important role.

3.5.1 An 8-State IP_3 Receptor Model

One of the earliest models of the IPR to incorporate sequential activation and inactivation by Ca^{2+} was that of De Young and Keizer (1992). Although a similar model by Atri et al. (1993) appeared at almost the same time it was the De Young-Keizer model that proved more popular, and so that is the one we discuss in detail here.

We assume that the IP_3 receptor consists of three equivalent and independent subunits, all of which must be in a conducting state before the receptor allows Ca^{2+} flux. Each subunit has an IP_3 binding site, an activating Ca^{2+} binding site, and an inactivating Ca^{2+} binding site, each of which can be either occupied or unoccupied, and thus each subunit can be in one of eight states. Each state of the subunit is labeled S_{ijk}, where i, j, and k are equal to 0 or 1, with a 0 indicating that the binding site is unoccupied and a 1 indicating that it is occupied. The first index refers to the IP_3 binding site, the second to the Ca^{2+} activation site, and the third to the Ca^{2+} inactivation site. This is illustrated in Fig. 8. Although a fully general model would include 24 rate constants, we make two simplifying assumptions. First, the rate constants are assumed to be independent of whether activating Ca^{2+} is bound or not. Second, the kinetics of Ca^{2+} activation are assumed to be independent of IP_3 binding and Ca^{2+} inactivation. This leaves only 10 rate constants, $k_1, \ldots, k_5$ and $k_{-1}, \ldots, k_{-5}$.

The fraction of subunits in the state S_{ijk} is denoted by x_{ijk}. The differential equations for these are based on mass-action kinetics, and thus, for example,

$$\frac{dx_{000}}{dt} = -(V_1 + V_2 + V_3) , \tag{26}$$

where

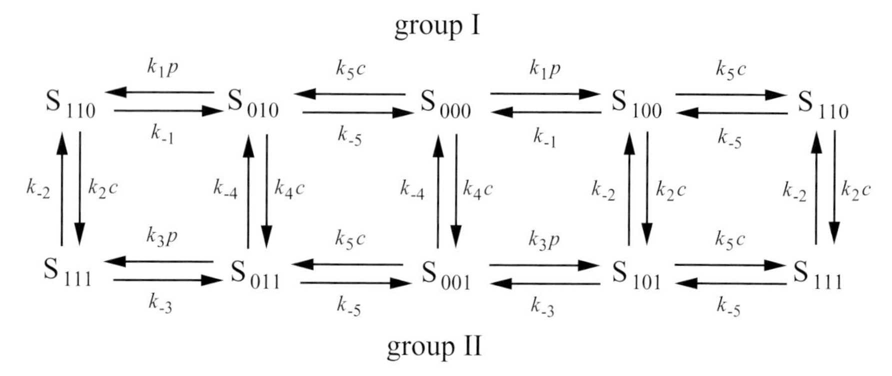

Fig. 8. The binding diagram for the IP_3 receptor model. Here, c denotes $[Ca^{2+}]$, and p denotes $[IP_3]$

$$V_1 = k_1 p x_{000} - k_{-1} x_{100} \,, \tag{27}$$

$$V_2 = k_4 c x_{000} - k_{-4} x_{001} \,, \tag{28}$$

$$V_3 = k_5 c x_{000} - k_{-5} x_{010} \,, \tag{29}$$

where p denotes $[IP_3]$ and c denotes $[Ca^{2+}]$. V_1 describes the rate at which IP_3 binds to and leaves the IP_3 binding site, V_2 describes the rate at which Ca^{2+} binds to and leaves the inactivating site, and similarly for V_3. The model assumes that the IP_3 receptor passes Ca^{2+} current only when three subunits are in the state S_{110} (i.e., with one IP_3 and one activating Ca^{2+} bound), and thus the open probability of the receptor is x_{110}^3.

In Fig. 9 we show the open probability of the IP_3 receptor as a function of $[Ca^{2+}]$, which is some of the experimental data upon which the model is based. Bezprozvanny et al. (1991) showed that this open probability is a bell-shaped function of $[Ca^{2+}]$. Thus, at low $[Ca^{2+}]$, an increase in $[Ca^{2+}]$ increases the open probability of the receptor, while at high $[Ca^{2+}]$ an increase in $[Ca^{2+}]$ decreases the open probability. Parameters in the model were chosen to obtain agreement with this steady-state data. The kinetic properties of the IP_3 receptor are equally important: the receptor is activated quickly by Ca^{2+}, but inactivated by Ca^{2+} on a slower time scale (i.e., $k_5 > k_2$).

3.5.2 A Model with Saturating Binding Rates

More recently it has become clear that the 8-state model has some serious flaws. Most importantly, it has been shown experimentally that the rate of Ca^{2+} binding to the IPR varies only over a single order of magnitude while the concentration of Ca^{2+} varies over four orders of magnitude. This can partially be seen in Fig. 7, although the limited time resolution obscures the issue; high time-resolution data from the laboratory of (Taylor, 1998; Swatton and Taylor, 2002; Taylor and Laude, 2002) makes the problem much clearer.

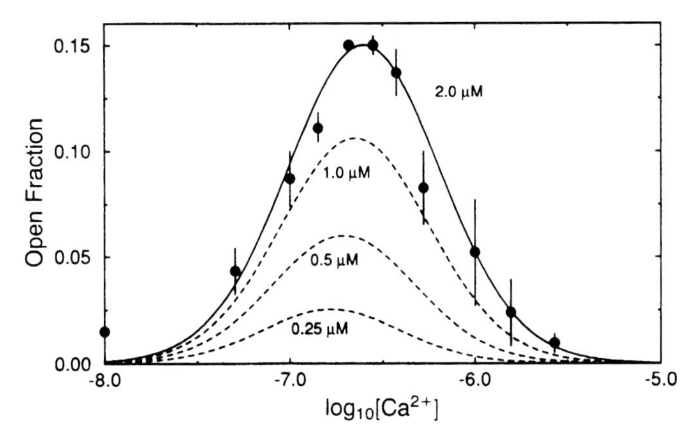

Fig. 9. The steady-state open probability of the IP_3 receptor, as a function of $[Ca^{2+}]$. The symbols are the experimental data of Bezprozvanny et al. (1991), and the smooth curves are from the 8-state receptor model (calculated at four different IP_3 concentrations) using the parameter values given in Table 1 (De Young and Keizer, 1992), Fig. 2A.)

Table 1. Parameters of the 8-state IPR model (De Young and Keizer, 1992)

$k_1 = 400\,\mu M^{-1}s^{-1}$	$k_{-1} = 52\,s^{-1}$
$k_2 = 0.2\,\mu M^{-1}s^{-1}$	$k_{-2} = 0.21\,s^{-1}$
$k_3 = 400\,\mu M^{-1}s^{-1}$	$k_{-3} = 377.2\,s^{-1}$
$k_4 = 0.2\,\mu M^{-1}s^{-1}$	$k_{-4} = 0.029\,s^{-1}$
$k_5 = 20\,\mu M^{-1}s^{-1}$	$k_{-5} = 1.64\,s^{-1}$

It thus follows that the binding of Ca^{2+} to the IPR cannot follow simple mass action kinetics, but must follow some kinetic scheme that allows for saturation of the binding rate.

Note how similar this is to the models of enzyme kinetics. There the saturation of the reaction rate showed how simple mass action kinetics was not appropriate to model enzyme reactions, thus leading to the development of the Michaelis-Menten approach. We follow a similar approach here to incorporate saturation of the binding rates (Fig. 10).

If we assume that $\tilde{A}$ and $\bar{A}$ are in instantaneous equilibrium we have

$$c\tilde{A} = L_1\bar{A}\,, \tag{30}$$

where $L_1 = l_{-1}/l_1$ and c denotes $[Ca^{2+}]$. Hence, letting $A = \bar{A} + \tilde{A}$, we have

$$\frac{dA}{dt} = (k_{-1} + l_{-2})I - \phi(c)A\,, \tag{31}$$

where

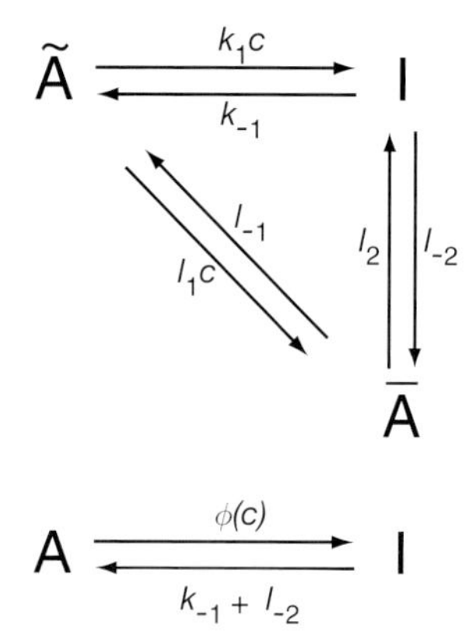

Fig. 10. Reaction diagram of a Ca^{2+}-dependent transition from state A to state I (c is Ca^{2+} concentration). If $\tilde{A}$ and $\bar{A}$ are in equilibrium, and if $A = \bar{A} + \tilde{A}$, then the upper reaction scheme is equivalent to the lower

$$\phi(c) = \frac{c(k_1 L_1 + l_2)}{c + L_1} \, . \tag{32}$$

Thus, this scheme is a simple way in which saturating binding kinetics can be incorporated into a model. The state $\bar{A}$ plays a role similar to that of the enzyme complex.

Using saturating binding schemes of this type, Sneyd and Dufour (2002) constructed a model of the IPR that was based on the qualitative models of Taylor and his colleagues, as well as being consistent with the scheme of Hajnóczky and Thomas (1997). In addition to the saturating binding rates, the main features of the model are

1. The IPR can be opened by IP_3 in the absence of Ca^{2+}, but with a lower conductance.
2. The IPR can be inactivated by Ca^{2+} in the absence of IP_3.
3. Once IP_3 is bound, the IPR can spontaneously inactivate, independently of Ca^{2+}.
4. Once IP_3 is bound, the IPR can also bind Ca^{2+} to activate the receptor. Thus there is an intrinsic competition between Ca^{2+}-mediated receptor activation and spontaneous inactivation.
5. Once the IPR is activated by Ca^{2+} binding, it can then be inactivated by binding of additional Ca^{2+}.
6. Binding of IP_3 and Ca^{2+} is sequential.

The binding scheme of the model is given in Fig. 11. Although it appears to contain a multiplicity of states, there are specific reasons for each one. The background structure is simple. Ignoring the various tildes, hats and primes, we see that a receptor, R, can bind Ca^{2+} and inactivate to state I_1, or it can bind IP_3 and open to state O. State O can then shut (state S), or bind Ca^{2+}

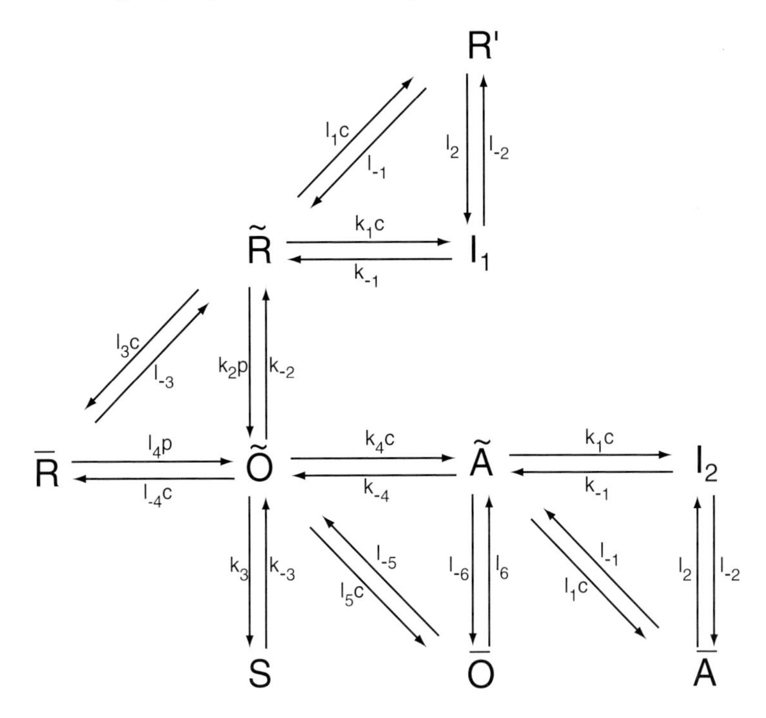

Fig. 11. The full IPR model. R-receptor; O-open; A-activated; S-shut; I-inactivated. c is $[Ca^{2+}]$, p is $[IP_3]$

and activate to state A. State A can then bind Ca^{2+} and inactivate to state I_2. This structure is clearer in Fig. 12.

States $\bar{R}$, $\bar{O}$, $\bar{A}$ and R' are used to give Ca^{2+}-dependent transitions that have saturable kinetics, as in the simple example of Fig. 10. These states will ultimately disappear, leaving behind only functions of c. Note that the inactivated states I_1 and I_2 both have Ca^{2+} bound to the same site, but I_2 also has IP_3 and one other Ca^{2+} ion bound. For simplicity we assume that the rate of Ca^{2+} binding to the inactivating site is independent of whether IP_3 is bound, or whether the receptor has been activated by Ca^{2+}. The $\bar{R}$, $\tilde{R}$, $\tilde{O}$ triangle models Ca^{2+}-dependent binding of IP_3; Ca^{2+} modulates the interconversion of the receptor between two states, each of which can bind IP_3 with different kinetics.

To derive the model equations we first define $K_i = k_{-i}/k_i$, and $L_i = l_{-i}/l_i$, for every appropriate integer i. We also let c and p denote $[Ca^{2+}]$ and $[IP_3]$ respectively. Then, assuming that the transitions $\tilde{R} \rightleftharpoons \bar{R}$, $\tilde{O} \rightleftharpoons \bar{O}$, $\tilde{A} \rightleftharpoons \bar{A}$ and $\tilde{R} \rightleftharpoons R'$ are fast and in instantaneous equilibrium, we get $c\tilde{R} = L_3\bar{R}$, $c\tilde{R} = L_1R'$, $c\tilde{O} = L_5\bar{O}$, and $c\tilde{A} = L_1\bar{A}$. We now define the new variables $R = \tilde{R} + \bar{R} + R'$, $O = \tilde{O} + \bar{O}$, $A = \tilde{A} + \bar{A}$. Then

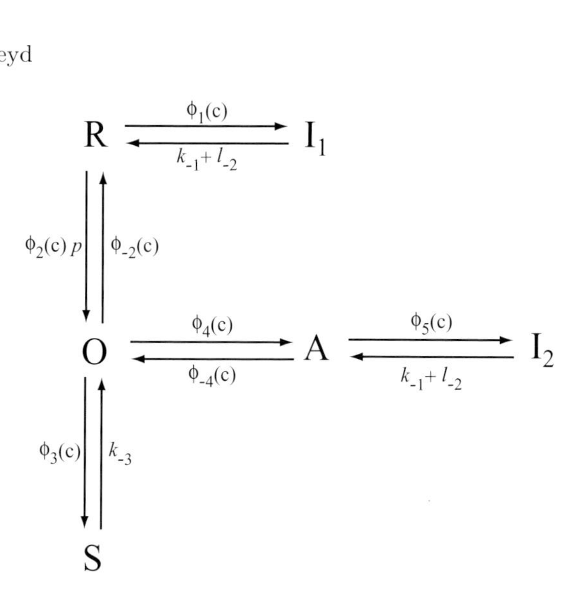

Fig. 12. Simplified diagram of the IPR model. Given the fast equilibria described in the text, this diagram is equivalent to that in Fig. 11

$$\frac{dR}{dt} = \phi_{-2}O - \phi_2 pR + (k_{-1} + l_{-2})I_1 - \phi_1 R \,, \tag{33}$$

$$\frac{dO}{dt} = \phi_2 pR - (\phi_{-2} + \phi_4 + \phi_3)O + \phi_{-4}A + k_{-3}S \,, \tag{34}$$

$$\frac{dA}{dt} = \phi_4 O - \phi_{-4}A - \phi_5 A + (k_{-1} + l_{-2})I_2 \,, \tag{35}$$

$$\frac{dI_1}{dt} = \phi_1 R - (k_{-1} + l_{-2})I_1 \,, \tag{36}$$

$$\frac{dI_2}{dt} = \phi_5 A - (k_{-1} + l_{-2})I_2 \,, \tag{37}$$

where

$$\phi_1(c) = \frac{(k_1 L_1 + l_2)c}{L_1 + c(1 + L_1/L_3)} \,, \tag{38}$$

$$\phi_2(c) = \frac{k_2 L_3 + l_4 c}{L_3 + c(1 + L_3/L_1)} \,, \tag{39}$$

$$\phi_{-2}(c) = \frac{k_{-2} + l_{-4}c}{1 + c/L_5} \,, \tag{40}$$

$$\phi_3(c) = \frac{k_3 L_5}{L_5 + c} \,, \tag{41}$$

$$\phi_4(c) = \frac{(k_4 L_5 + l_6)c}{L_5 + c} \,, \tag{42}$$

$$\phi_{-4}(c) = \frac{L_1(k_{-4} + l_{-6})}{L_1 + c} \,, \tag{43}$$

$$\phi_5(c) = \frac{(k_1 L_1 + l_2)c}{L_1 + c} , \tag{44}$$

and where $R+O+A+S+I_1+I_2=1$. Thus, given the fast equilibria above, Fig. 11 is equivalent to Fig. 12.

The parameters were determined by fitting to the experimental data shown in Fig. 7 and are shown in Table 2. Although the model is overparameterised (at least for the purposes of fitting to the data in Fig. 7), it does a better job than other models in the literature (Sneyd et al., 2004b). We assume that the IPR consists of four identical and independent subunits, and that it allows Ca^{2+} current when all four subunits are in either the O or the A state. We also assume that the more subunits in the A state, the greater the conductance. One simple way of summarising this is to write the open probability, P_O, as

$$P_O = (a_1 O + a_2 A)^4 , \tag{45}$$

for some constants a_1 and a_2. In the original model $a_1 = 0.1$ and $a_2 = 0.9$, but these values could be changed without appreciably changing the fit or the model's behaviour.

Table 2. Parameters of the IPR model with saturating binding rates (Sneyd and Dufour, 2002)

$k_1 = 0.64\,\mathrm{s}^{-1}\mu\mathrm{M}^{-1}$	$k_{-1} = 0.04\,\mathrm{s}^{-1}$
$k_2 = 37.4\,\mu\mathrm{M}^{-1}\mathrm{s}^{-1}$	$k_{-2} = 1.4\,\mathrm{s}^{-1}$
$k_3 = 0.11\,\mu\mathrm{M}^{-1}\mathrm{s}^{-1}$	$k_{-3} = 29.8\,\mathrm{s}^{-1}$
$k_4 = 4\,\mu\mathrm{M}^{-1}\mathrm{s}^{-1}$	$k_{-4} = 0.54\,\mathrm{s}^{-1}$
$L_1 = 0.12\,\mu\mathrm{M}$	$L_3 = 0.025\,\mu\mathrm{M}$
$L_5 = 54.7\,\mu\mathrm{M}$	$l_2 = 1.7\,\mathrm{s}^{-1}$
$l_4 = 1.7\,\mathrm{s}^{-1}\mu\mathrm{M}^{-1}$	$l_6 = 4707\,\mathrm{s}^{-1}$
$l_{-2} = 0.8\,\mathrm{s}^{-1}$	$l_{-4} = 2.5\,\mu\mathrm{M}^{-1}\mathrm{s}^{-1}$
$l_{-6} = 11.4\,\mathrm{s}^{-1}$	

3.6 Putting the Pieces Together: a Model of Calcium Oscillations

We now have constructed models for most of the fluxes that are important for the control of Ca^{2+}. To put them all together into a model of Ca^{2+} oscillations is a simple task. You just choose your favorite model of the IPR (and RyR if you wish to include that flux also), your favourite models of the ATPases and the influx, and put them all into (1) and (2). A large selection of such models is discussed in Falcke (2004).

For example, if we

1. ignore RyR and mitochondrial fluxes,
2. use the saturating binding model for the IPR,
3. use a four-state Markov model for the SERCA pump, with the transport of only a single Ca^{2+} ion for each pump cycle,
4. use a simple Hill function for the plasma membrane pump, and
5. assume that buffering is fast and linear (see Sect. 5)

we get

$$\frac{dc}{dt} = (k_f P_O + J_{\text{er}})(c_e - c) - J_{\text{serca}} + J_{\text{in}} - J_{\text{pm}} \tag{46}$$

$$\frac{dc_e}{dt} = \gamma[J_{\text{serca}} - (k_f P_O + J_{\text{er}})(c_e - c)], \tag{47}$$

where

$$J_{\text{serca}} = \frac{c - \alpha_1 c_e}{\alpha_2 + \alpha_3 c + \alpha_4 c_e + \alpha_5 c c_e} \tag{48}$$

$$P_O = (0.1O + 0.9A)^4 \tag{49}$$

$$J_{\text{pm}} = \frac{V_p c^2}{K_p^2 + c^2} \tag{50}$$

$$J_{\text{er}} = \text{constant} \tag{51}$$

$$J_{\text{in}} = a_1 + a_2 p\ . \tag{52}$$

Note how the flux through the IPR is assumed to be proportional to the concentration difference between the ER and the cytoplasm. The constant J_{er} models a constant leak from the ER that is necessary to balance the ATPase pump at steady state. Equations (46) and (47) are coupled to the (33)–(44) describing the IPR. We have used two different pump models to emphasise the fact that it is possible to mix and match the individual models for the various fluxes. Since there are five parameters describing the SERCA ATPase pump, and only two parameters describing the plasma membrane ATPase pump, it seems likely that the SERCA pump is overparameterised. If we were fitting the pump models to data this would indeed be a concern. Here, however, we merely wish to use some function that can be justified by a mechanistic model and one for which the parameters can be adjusted to give reasonable agreement with experimental data. The parameters used here for the SERCA model are based on the work of Favre et al. (1996), while the simpler model for the plasma membrane pump is based on Lytton et al. (1992). All the parameter values are given in Table 3.

Since p corresponds to the concentration of IP_3, and thus, indirectly, the agonist concentration, we study the behavior of this model as p varies. The bifurcation diagram is shown in Fig. 13A. As p increases the steady state increases also (because J_{in} increases with p), and oscillations occur for a

Table 3. Parameter values of the model of Ca^{2+} oscillations (46)–(52)

$\alpha_1 = 10^{-4}$	$\alpha_2 = 0.007\,\text{s}$
$\alpha_3 = 0.06\,\mu\text{M}^{-1}\,\text{s}$	$\alpha_4 = 0.0014\,\mu\text{M}^{-1}\text{s}$
$\alpha_5 = 0.007\,\mu\text{M}^{-2}\,\text{s}$	$J_{\text{er}} = 0.002\,\text{s}^{-1}$
$V_p = 28\,\mu\text{M}\,\text{s}^{-1}$	$K_p = 0.425\,\mu\text{M}$
$a_1 = 0.03\,\mu\text{M}\,\text{s}^{-1}$	$a_2 = 0.2\,\text{s}^{-1}$
$k_f = 0.96\,\text{s}^{-1}$	$\gamma = 5.5$
$\delta = 0.1$	

range of intermediate values of p. Typical oscillations for two different values of p are shown in Fig. 13B and C. As p increases the oscillation frequency increases.

Given the variety of models of the IPR, the RyR, the ATPases, and the other fluxes involved in Ca^{2+} dynamics, it is very important to think about what a model for Ca^{2+} oscillations is telling us. For instance, given just about any combination of IPR and ATPase models, we can construct a whole-cell model that exhibits Ca^{2+} oscillations of approximately the correct form. The oscillations in Fig. 13 are a reasonably accurate description of Ca^{2+} oscillations in pancreatic acinar cells (the cell type for which that model was originally designed); however, the parameters could be adjusted, practically indefinitely, to obtain oscillations of a wide variety of other shapes and properties to describe the behavior in other cell types. Therefore, the fact that the model exhibits Ca^{2+} oscillations with reasonable properties tells us very little about the underlying mechanisms. We may have written down one possible mechanism (out of many) but a great deal more work is needed before we can conclude anything about mechanisms in actual cells.

3.7 IP_3 Dynamics

In the above model Ca^{2+} oscillations occur at a constant concentration of IP_3. Thus the role of IP_3 is merely to open the IPR; once the receptor is opened, Ca^{2+} feedback takes over and the period of the oscillations is then controlled by the kinetics of Ca^{2+} feedback on the IPR as well as the interaction of the membrane and ER fluxes.

However, this is certainly an over-simplification. It is known that the rate of production of IP_3 is dependent on Ca^{2+}, and it is known that, in some cell types, oscillations in $[Ca^{2+}]$ are accompanied by oscillations in $[IP_3]$ (Hirose et al., 1999; Nash et al., 2001; Young et al., 2003). What is not yet clear is whether oscillations in $[IP_3]$ are *necessary* for Ca^{2+} oscillations, as the former could merely be a passive follower of the latter. Experimental evidence is neither consistent nor conclusive.

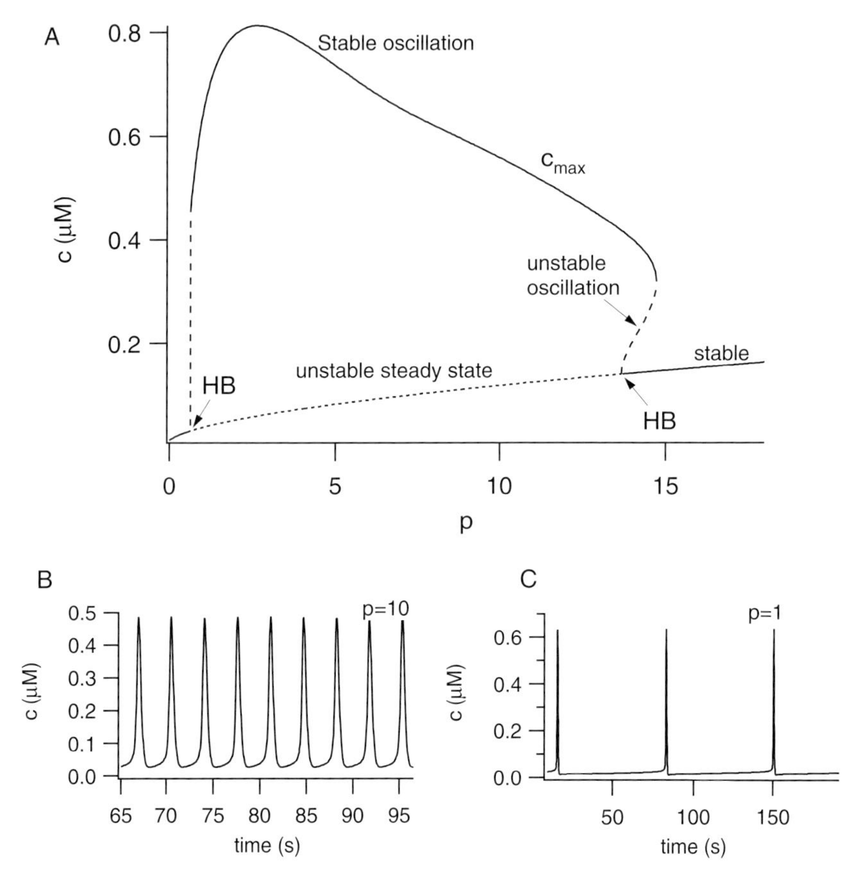

Fig. 13. (**A**) Simplified bifurcation diagram of the model of calcium oscillations (46)–(52). We plot the steady state and the maximum of the oscillation, with a *solid line* denoting stability and a *dashed line* instability. HB denotes a Hopf bifurcation. Although it appears as if the branch of oscillations shown here begins at the lower Hopf bifurcation this is actually not so. The actual diagram is much more complicated in that region and not given in full here. (**B**) and (**C**) typical oscillations in the model for two different values of p

There are a number of models that incorporate the Ca^{2+} dependence of IP_3 production and degradation. Early models were those of Meyer and Stryer (1988), Swillens and Mercan (1990), De Young and Keizer (1992) and Cuthbertson and Chay (1991), more recent models have been constructed by Shen and Larter (1995); Dupont and Erneux (1997) and Houart et al. (1999).

There are two principal ways in which Ca^{2+} can influence the concentration of IP_3 and these are sketched in Fig. 14. Firstly, activation of PLC by Ca^{2+} is known to occur in many cell types; this forms the basis of the models of Meyer and Stryer (1988) and De Young and Keizer (1992). Secondly, IP_3 is

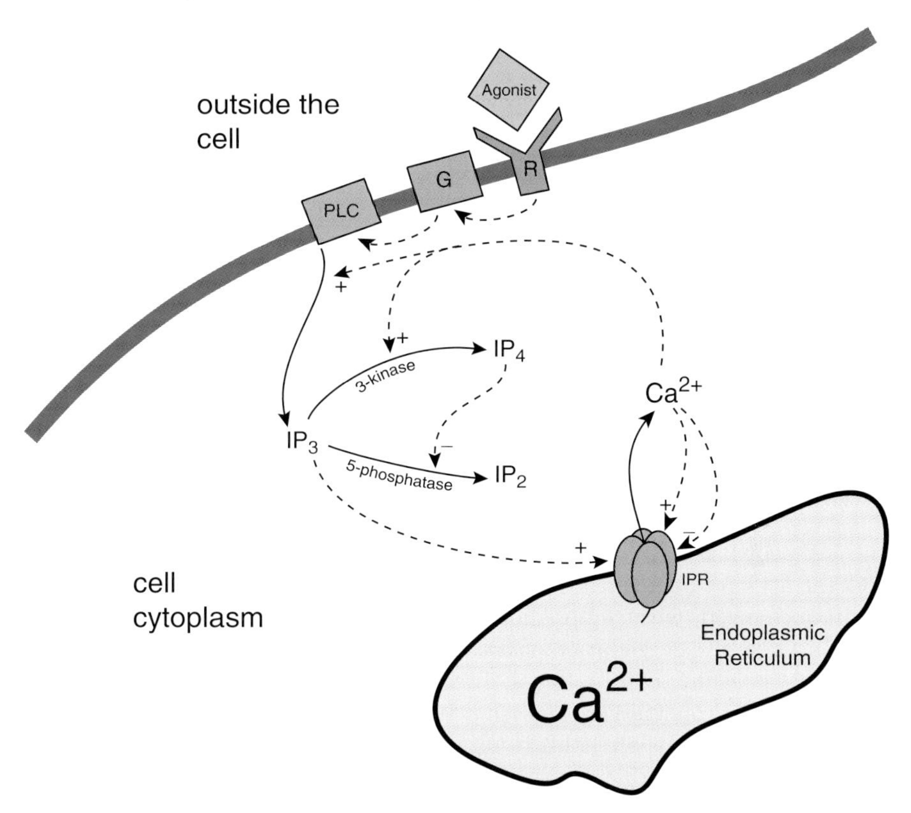

Fig. 14. Schematic diagram of some of the interactions between Ca^{2+} and IP_3 dynamics. Calcium can activate PLC, leading to an increase in the rate of production of IP_3, and it can also increase the rate at which IP_3 is phosphorylated by the 3-kinase. The end product of phosphorylation by the 3-kinase, IP_4, acts as a competitive inhibitor of dephosphorylation of IP_3 by the 5-phosphatase. Not all of these feedbacks are significant in every cell type

degraded in two principal ways, each of which is subject to feedback regulation; it is dephosphorylated by IP_3 5-phosphatase to inositol 1,4-bisphosphate (IP_2), or phosphorylated by IP_3 3-kinase to inositol 1,3,4,5-tetrakisphosphate (IP_4). However, since IP_4 is also a substrate for the 5-phosphatase it acts as a competitive inhibitor of IP_3 dephosphorylation. In addition, Ca^{2+} (in the form of a Ca^{2+}-calmodulin complex) can enhance the activity of the 3-kinase.

Intuitively it seems clear that such an intricate array of feedbacks could easily give rise to interdependent oscillations in IP_3 and Ca^{2+}, and the models mentioned above have confirmed that this intuitive expectation is correct. However, few of those models have then gone on to make predictions and test them experimentally. One notable exception is that of Dupont et al. (2003). They studied agonist-induced oscillations in hepatocytes, in which activation of PLC by Ca^{2+} does not appear to play a significant role. Thus, their model concentrated on the two pathways by which IP_3 is degraded, the 3-kinase and the 5-phosphatase.

In hepatocytes, oscillations of Ca^{2+} and IP_3 occur together. Each Ca^{2+} spike causes an increase in the rate of phosphorylation of IP_3, and thus a decline in $[IP_3]$, with a consequent decline in $[Ca^{2+}]$. However, although both species oscillate, it is not clear that oscillations in $[IP_3]$ are *necessary*. Dupont et al. first took a model (Dupont and Erneux, 1997) in which Ca^{2+} influences the rate of the 3-kinase. This model predicted that, if the rate of the 5-phosphatase is increased by a factor of 25, and the rate of IP_3 production also increased, oscillations with the same qualitative properties persist, with only a slight decrease in frequency. Under these conditions, very little IP_3 is being degraded by the 3-kinase, and thus Ca^{2+} feedback on the 3-kinase cannot be playing a major role. In other words, these oscillations arise via the feedback of Ca^{2+} on the IPR.

Dupont et al. then devised an experimental test of this model prediction. They increased both the rate of the 5-phosphatase (by direct microinjection of the 5-phosphatase), as well as the rate of production of IP_3 (by increasing the concentration of agonist), and observed that the oscillations remained essentially unchanged, with the predicted decrease in frequency. Thus, under conditions where the degradation of IP_3 is principally through the 5-phosphatase pathway (in which case it is not directly affected by $[Ca^{2+}]$) there is little change in the properties of the oscillations. The conclusion is that, in hepatocytes, the oscillations are not a result of the interplay between $[Ca^{2+}]$ and the IP_3 dynamics, but rather a result of Ca^{2+} feedback on the IPR. This work is an elegant example of how a model should be used.

4 Calcium Waves

In some cell types, Ca^{2+} oscillations occur practically uniformly across the cell. In such a situation, measurement of the Ca^{2+} concentration at any point of the cell gives the same time course, and a well-mixed model is thus an appropriate one. More often, however, each oscillation takes the form of a wave across the cell; these intracellular "oscillations" are actually periodic intracellular waves. To model and understand such spatially distributed behavior, inclusion of Ca^{2+} diffusion is vital. Furthermore, to study such things as spiral waves of Ca^{2+} (Fig. 3) a partial differential equation model is again necessary.

It is usually assumed that Ca^{2+} diffuses with constant diffusion coefficient D_c, and that the cytoplasm is isotropic and homogeneous. Exceptions to the assumption of homogeneity are not uncommon, but exceptions to the isotropic assumption are rare. We also assume that c and c_e coexist at every point in space. Obviously this is not what really occurs. However, if the ER is "smeared out" sufficiently, it is reasonable to assume that every point in space is close enough to both the ER and the cytoplasm to allow for such a simplified model. In this case the reaction-diffusion equation for Ca^{2+} is

$$-\frac{\partial c}{\partial t} = D_c \nabla^2 c + J_{\text{IPR}} + J_{\text{RyR}} - J_{\text{serca}} - J_{\text{on}} + J_{\text{off}} + J_{\text{uni}} - J_{\text{mito}} \,. \tag{53}$$

Note how the fluxes J_{in} and J_{pm} are omitted as they are fluxes only on the boundary of the domain and thus must be treated differently.

Modeling the concentration of Ca^{2+} in the ER is considerably less straightforward. For instance, it is not known how well Ca^{2+} diffuses in the ER, or whether the tortuosity of the ER plays any role in determining an effective diffusion coefficient for ER Ca^{2+}. In general, modelers have assumed either that Ca^{2+} does not diffuse at all in the ER, or that it does so with a restricted diffusion coefficient, D_e. In either case

$$\frac{\partial c_e}{\partial t} = D_e \nabla^2 c_e - \gamma(J_{\text{IPR}} + J_{\text{RyR}} - J_{\text{serca}}) - J_{\text{on,e}} + J_{\text{off,e}} \,, \tag{54}$$

with D_e varying from 0 up to D_c depending on the exact assumptions.

The boundary conditions must be approached with caution. The reason for this is that cells are inherently three dimensional objects. When modeled in three dimensions the boundary conditions are clear. At the boundary of the cell (i.e., the plasma membrane) the flux into the cell is $J_{\text{in}} - J_{\text{pm}}$. Thus

$$D_c \nabla c \cdot n = J_{\text{in}} - J_{\text{pm}} \tag{55}$$

$$\nabla c_e \cdot n = 0 \,, \tag{56}$$

where n is normal to the boundary. Note that we assume that the ER cannot communicate directly with the outside of the cell (although this assumption, as with almost everything in biology, is not itself without controversy).

However, when the equations are solved in one or two dimensions this simplification introduces spurious boundaries which are not so easily dealt with. It is left as an exercise to show that, in the limit as, say, a two dimensional region becomes infinitely thin, these spurious boundaries have zero-flux conditions and that the fluxes J_{in} and J_{pm} just appear as additional fluxes on the right hand side of the reaction-diffusion equation.

For example, if we wished to solve the equations on the domain $0 \le x \le 20$ with no diffusion of Ca^{2+} in the ER, then we would impose the boundary conditions $\frac{\partial c}{\partial x} = 0$ at $x = 20$ and $x = 0$, while at each internal point we would have

$$\frac{\partial c}{\partial t} = D_c \nabla^2 c + J_{\text{IPR}} + J_{\text{in}} - J_{\text{pm}} + J_{\text{RyR}} - J_{\text{serca}} - J_{\text{on}} + J_{\text{off}} \tag{57}$$

$$\frac{\partial c_e}{\partial t} = \gamma(J_{\text{serca}} - J_{\text{IPR}} - J_{\text{RyR}}) - J_{\text{on,e}} + J_{\text{off,e}} \,. \tag{58}$$

We have omitted the mitochondrial fluxes for simplicity.

There are a number of ways to study reaction-diffusion models of this type, but the two most common (at least in the study of Ca^{2+} waves) are, firstly, numerical simulation or secondly, bifurcation analysis of the traveling wave equations.

4.1 Simulation of Spiral Waves in *Xenopus*

One common experimental procedure for initiating waves in *Xenopus* oocytes is to photorelease a bolus of IP_3 inside the cell and observe the subsequent Ca^{2+} activity (Lechleiter and Clapham, 1992). After sufficient time, Ca^{2+} wave activity disappears as IP_3 is degraded, but in the short term, the observed Ca^{2+} activity is the result of Ca^{2+} diffusion and IP_3 diffusion. Another technique is to release IP_3S_3, a nonhydrolyzable analogue of IP_3, which has a similar effect on IP_3 receptors but is not degraded by the cell. In this case, after sufficient time has passed, the IP_3S_3 is at a constant concentration in all parts of the cell.

When a bolus of IP_3S_3 is released in the middle of the domain, it causes the release of a large amount of Ca^{2+} at the site of the bolus. The IP_3S_3 then diffuses across the cell, releasing Ca^{2+} in the process. Activation of IP_3 receptors by the released Ca^{2+} can lead to periodic Ca^{2+} release from the stores, and the diffusion of Ca^{2+} between IP_3 receptors serves to stabilize the waves, giving regular periodic traveling waves. These periodic waves are the spatial analogues of the oscillations seen in the temporal model, and arise from the underlying oscillatory kinetics. If the steady $[IP_3S_3]$ is in the appropriate range (see, for example, Fig. 13, which shows that limit cycles exist for $[IP_3]$ in some intermediate range) over the entire cell, every part of the cell cytoplasm is in an oscillatory state. It follows from the standard theory of reaction–diffusion systems with oscillatory kinetics (see, for example, Kopell and Howard, 1973; Duffy et al., 1980; Neu, 1979; Murray, 1989) that periodic and spiral waves can exist for these values of $[IP_3S_3]$. When IP_3, rather than IP_3S_3, is released, the wave activity lasts for only a short time. These waves were simulated by Atri et al. (1993) who showed that when the wave front is broken, a spiral wave of Ca^{2+} often forms (Fig. 3B). Depending on the initial conditions, these spiral waves can be stable or unstable. In the unstable case, the branches of the spiral can intersect themselves and cause breakup of the spiral, in which case a region of complex patterning emerges in which there is no clear spatial structure (McKenzie and Sneyd, 1998).

A more detailed understanding of the stability of spiral calcium waves has been developed by Falcke et al. (1999, 2000), who showed that an increased rate of Ca^{2+} release from the mitochondria can dramatically change the kinds of waves observed, and that, in extreme cases, the cytoplasm can be made bistable, with an additional steady state with a high resting $[Ca^{2+}]$. When this happens the spiral waves become unstable, an instability that appears to be related to a gap in the dispersion curve.

4.2 Traveling Wave Equations and Bifurcation Analysis

The second principal way in which Ca^{2+} waves are studied is the analysis of the traveling wave equations. This is most useful for the study of waves in one spatial dimension. If we introduce the traveling wave variable $\xi = x + st$, where s is the wave speed, we can write (58) as the pair of equations

$$c' = d \tag{59}$$

$$D_c d' = sd - \sum J \,, \tag{60}$$

where $\sum J$ denotes all the various fluxes on the right hand side of (58), and where a prime denotes differentiation with respect to ξ. Thus a single reaction-diffusion equation is converted to two ordinary differential equation, resulting in considerable simplification. In general these two equations will be coupled to the other equations for c_e, p, and the states of the various receptors. When the equations are written as a function of the variable ξ we shall call them the traveling wave equations.

Traveling pulses, traveling fronts, and periodic waves correspond to, respectively, homoclinic orbits, heteroclinic orbits and limit cycles in the traveling wave equations. Thus, by studying the bifurcations in the traveling wave equations we can gain considerable insight into what kinds of waves exist in the model, and for which parameter values. However, such an approach does not readily give information about the *stability* of the wave solutions of the original reaction-diffusion equations. That is much more difficult to determine. Numerical simulation of the reaction-diffusion equations can begin to address the question of stability, although care must be used.

We shall briefly illustrate the method using the model of Sect. 3.6. Adding the diffusion of c (but not of c_e) to that model gives

$$\frac{\partial c}{\partial t} = D_c \frac{\partial^2 c}{\partial x^2} + (k_f P_O + J_{\mathrm{er}})(c_e - c) - J_{\mathrm{serca}} + J_{\mathrm{in}} - J_{\mathrm{pm}} \tag{61}$$

$$\frac{\partial c_e}{\partial t} = \gamma [J_{\mathrm{serca}} - (k_f P_O + J_{\mathrm{er}})(c_e - c)] \,, \tag{62}$$

and these two equations are coupled, as before, to the five equations of the 6-state IPR model (Sect. 3.5.2). Rewriting these equations in the traveling wave variable then gives

$$c' = d \tag{63}$$

$$D_c d' = sd - (k_f P_O + J_{\mathrm{er}})(c_e - c) + J_{\mathrm{serca}} - J_{\mathrm{in}} + J_{\mathrm{pm}} \tag{64}$$

$$sc_e' = \gamma [J_{\mathrm{serca}} - (k_f P_O + J_{\mathrm{er}})(c_e - c)] \tag{65}$$

coupled to the five receptor equations. We note that, given the units of the parameters in Table 3, the speed s has the units $\mu\mathrm{m\,s^{-1}}$.

We now do a two-parameter bifurcation analysis of these equations, using s and p as the bifurcation parameters (Fig. 15). The behavior as $s \to \infty$ is just that of the model in the absence of diffusion, as expected from the general theory (Maginu, 1985). Thus, for large values of s there are two Hopf bifurcations, with a branch of periodic solutions for intermediate values of p.

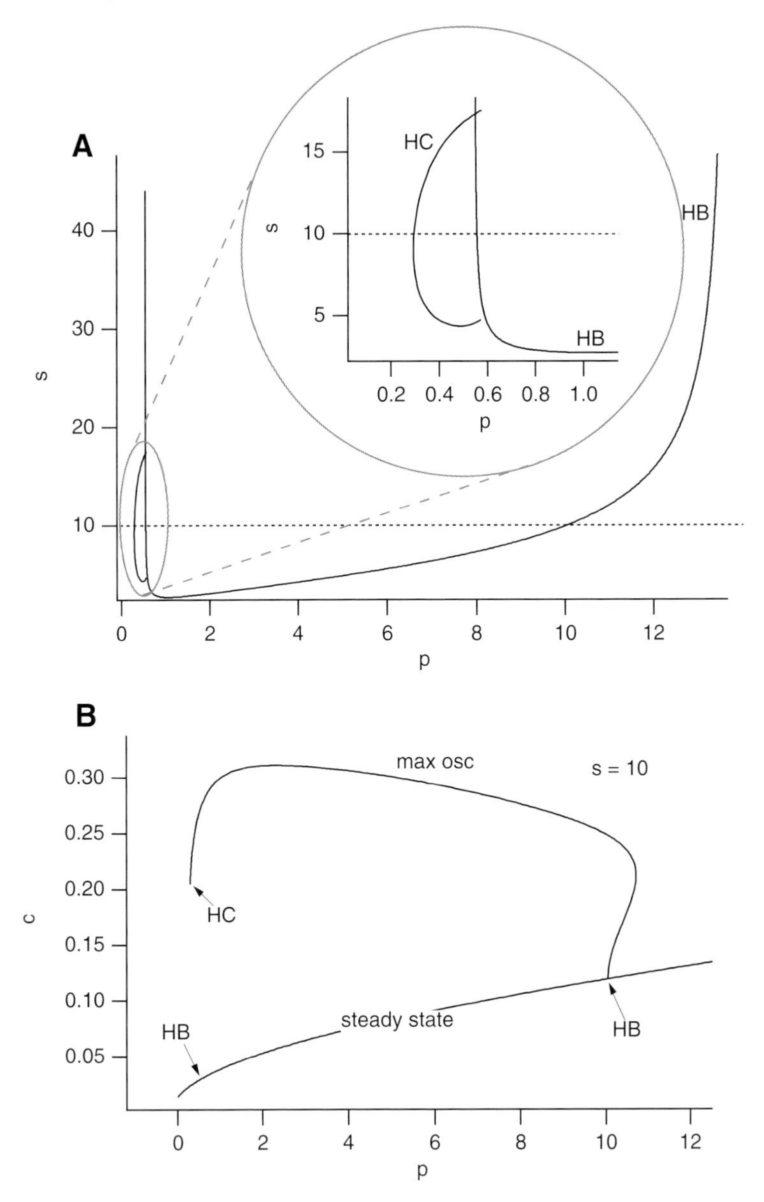

Fig. 15. (**A**) Two-parameter bifurcation diagram of the traveling wave equations of the model of Ca^{2+} wave propagation. HC – homoclinic bifurcation; HB – Hopf bifurcation. The inset shows a blowup of the branch of homoclinic bifurcations. (**B**) bifurcation diagram obtained by taking a cross-section at $s = 10$, as shown by the *dotted line* in panel A. Both panels were computed using the parameter values in Table 3, and using $D_c = 25\ \mu m^2 s^{-1}$

If we track these Hopf bifurcations in the s, p plane they form a U-shaped curve (Fig. 15A). To find homoclinic bifurcations corresponding to traveling waves, we take a cross-section of this diagram for a constant value of $s = 10$, i.e., for that value of s we plot the bifurcation diagram of c against p (just as in Fig. 13). The result is shown in Fig. 15B. The two Hopf bifurcations correspond to where the U-shaped curve in panel A intersects the dotted line at $s = 10$. Tracking the branch of periodic orbits that comes out of the upper Hopf bifurcation gives a branch that ends in a homoclinic orbit. If we then track this homoclinic orbit in the s, p plane we get the C-shaped curve of homoclinics shown in the inset to panel A. More accurately, we can only track branches of very high period (10,000 in this case), so the curve labeled HC in panel A is really just a branch of high period orbits. However, tracking the branch of orbits of period 1,000, or of period 30,000 gives an identical result and thus we can be relatively confident that we are tracking a homoclinic orbit.

It turns out that this branch of homoclinic orbits is the one that gives stable traveling wave solutions of the original reaction-diffusion equation (when solved as a partial differential equation, not in the traveling wave coordinates). This is easily checked numerically by direct simulation; if one fixes $p = 0.5$, gives a large Ca^{2+} stimulus at one end of the (one-dimensional) domain, and observes the traveling wave that results, one finds that it travels at approximately speed 15 $\mu m\,s^{-1}$ and has the same shape as the homoclinic orbit on that C-shaped branch.

A number of interesting results appear immediately. Firstly, the branch of homoclinic orbits predicts traveling waves with speeds of around 10–17 $\mu m\,s^{-1}$, exactly in the physiological range. Secondly, as p increases, so does the wave speed, something that has been observed experimentally. Thirdly, the ends of the C-shaped homoclinic branch are puzzling. Numerical results clearly indicate that the curve of homoclinic bifurcations passes through the curve of Hopf bifurcations, but this is difficult to justify theoretically. Although this end of the C-shaped homoclinic branch is not at all understood it is physiologically important as it corresponds to the waves that would be observed as p is slowly increased. At some point there is a transition from single traveling pulses to periodic waves, but exactly how that transition works is not well understood. Finally, more detailed investigation shows a vastly more complicated bifurcation structure which has yet to be fully elucidated.

The basic structure of the bifurcation diagram shown in Fig. 15, with a C-shaped branch of homoclinic orbits and a U-shaped branch of Hopf bifurcations, seems to be generic to models of excitable systems. The FitzHugh-Nagumo model has the same basic structure, as does the Hodgkin-Huxley model. All models of Ca^{2+} wave propagation (at least all the ones for which this question has been studied) also have the same basic C-U structure in the s, p plane. Thus, although studies of such systems are still in their infancy, they have great potential.

5 Calcium Buffering

Calcium is heavily buffered in all cells, with at least 99% (and often more) of the available Ca^{2+} bound to large Ca^{2+}-binding proteins of which there are about 200 encoded by the human genome (Carafoli et al., 2001). For instance, calsequestrin and calreticulin are major calcium buffers in the endoplasmic and sarcoplasmic reticulum, while in the cytoplasm Ca^{2+} is bound to calbindin, calretinin and parvalbumin, among many others. Calcium pumps and exchangers and the plasma membrane itself are also major Ca^{2+} buffers. In essence, a free Ca^{2+} ion in solution in the cytoplasm cannot do very much, or go very far, before it is bound to one thing or another.

The basic reaction for calcium buffering can be represented by the reaction

$$\mathrm{P} + \mathrm{Ca}^{2+} \underset{k_-}{\overset{k_+}{\rightleftharpoons}} \mathrm{B} , \tag{66}$$

where P is the buffering protein and B is buffered calcium. If we let b denote the concentration of buffer with Ca^{2+} bound, and c the concentration of free Ca^{2+}, then a simple model of calcium buffering is

$$\frac{\partial c}{\partial t} = D_c \nabla^2 c + f(c) + k_- b - k_+ c(b_t - b) , \tag{67}$$

$$\frac{\partial b}{\partial t} = D_b \nabla^2 b - k_- b + k_+ c(b_t - b) , \tag{68}$$

where k_- is the rate of Ca^{2+} release from the buffer, k_+ is the rate of Ca^{2+} uptake by the buffer, b_t is the total buffer concentration, and $f(c)$ denotes all the other reactions involving free Ca^{2+} (for example, release from the IP_3 receptors, reuptake by pumps, etc.).

5.1 Fast Buffers or Excess Buffers

If the buffer has fast kinetics, its effect on the intracellular Ca^{2+} dynamics can be analyzed simply. For if k_- and k_+ are large compared to the time constant of calcium reaction, then we take b to be in the quasi-steady state

$$k_- b - k_+ c(b_t - b) = 0 , \tag{69}$$

and so

$$b = \frac{b_t c}{K + c} , \tag{70}$$

where $K = k_-/k_+$. It follows that

$$\frac{\partial c}{\partial t} + \frac{\partial b}{\partial t} = (1 + \theta)\frac{\partial c}{\partial t} , \tag{71}$$

where

$$\theta(c) = \frac{b_t K}{(K+c)^2} \; . \tag{72}$$

Combining (71) with (67) and (68), we obtain

$$\frac{\partial c}{\partial t} = \frac{1}{1+\theta(c)} \left(\nabla^2 \left(D_c c + D_b b_t \frac{c}{K+c} \right) + f(c) \right) \tag{73}$$

$$= \frac{D_c + D_b\theta(c)}{1+\theta(c)} \nabla^2 c - \frac{2D_b\theta(c)}{(K+c)(1+\theta(c))} |\nabla c|^2 + \frac{f(c)}{1+\theta(c)} \; . \tag{74}$$

Note that we are assuming that b_t is a constant, and doesn't vary in either space or time.

We see that nonlinear buffering changes the model significantly. In particular, Ca^{2+} obeys a nonlinear diffusion–advection equation, where the advection is the result of Ca^{2+} transport by a mobile buffer (Wagner and Keizer, 1994). The effective diffusion coefficient

$$D_{\text{eff}} = \frac{D_c + D_b\theta(c)}{1+\theta(c)} \tag{75}$$

is a convex linear combination of the two diffusion coefficients D_c and D_b, and so lies somewhere between the two. Since buffers are large molecules, $D_{\text{eff}} < D_c$. If the buffer is not mobile, i.e., $D_b = 0$, then (74) reverts to a reaction–diffusion equation. Also, when Ca^{2+} gradients are small, the nonlinear advective term can be ignored (Irving et al., 1990). Finally, the buffering also affects the qualitative nature of the nonlinear reaction term, $f(c)$, which is divided by $1+\theta(c)$. This may change many properties of the model, including oscillatory behavior and the nature of wave propagation.

If the buffer is not only fast, but also of low affinity, so that $K \gg c$, it follows that

$$b = \frac{b_t c}{K} \tag{76}$$

in which case

$$\theta = \frac{b_t}{K} \; , \tag{77}$$

a constant. Thus D_{eff} is constant also.

Very commonly it is assumed that the buffer has fast kinetics, is immobile, and has a low affinity. With these assumptions we get the simplest possible way to model Ca^{2+} buffers (short of not including them at all) in which

$$\frac{\partial c}{\partial t} = \frac{1}{1+\theta} (D_c \nabla^2 c + f(c)) \; . \tag{78}$$

Both the diffusion coefficient and the fluxes are merely scaled by the constant factor $1/(1+\theta)$; each flux in the model can then be interpreted as an *effective* flux, i.e., that fraction of the actual flux that contributes towards a change in free Ca^{2+} concentration.

In 1986 Neher observed that if the buffer is present in large excess then $b_t - b \approx b_t$, in which case the buffering equations become linear, the so-called excess buffering approximation:

$$\frac{\partial c}{\partial t} = D_c \nabla^2 c + f(c) + k_- b - k_+ c b_t \ , \quad (79)$$

$$\frac{\partial b}{\partial t} = D_b \nabla^2 b - k_- b + k_+ c b_t \ . \quad (80)$$

If we then assume the buffers are fast we recover (76) and thus (78). In other words, the simple approach to buffering given in (78) can be obtained in two ways; either by assuming a low affinity buffer or by assuming that the buffer is present in excess. It is intuitively clear why these two approximations lead to the same result – in either case the binding of Ca^{2+} does little to change the fraction of unbound buffer.

Typical parameter values for three different buffers are given in Table 4. BAPTA is a fast, high-affinity buffer, and EGTA is a slow high-affinity buffer, both of which are used as exogeneous Ca^{2+} buffers in experimental work. Parameters for a typical endogenous buffer are also included.

Many studies of Ca^{2+} buffering have addressed the problem of the steady-state Ca^{2+} distribution that arises from the flux of Ca^{2+} through a single open channel (Naraghi and Neher, 1997; Stern 1992; Smith, 1996; Smith et al., 1996, 2001; Falcke, 2003). Such studies are motivated by the fact that, when a channel opens, the steady-state Ca^{2+} distribution close to the mouth of the channel is reached within microseconds. In that context, Smith et al. (2001) have derived both the rapid buffering approximation and the excess buffering approximation as asymptotic solutions of the original equations.

Table 4. Typical parameter values for three different buffers, taken from Smith et al. (2001). BAPTA and EGTA are commonly used as exogenous buffers in experimental work, while Endog refers to a typical endogenous buffer. Typically $b_t = 100\,\mu M$ for an endogenous buffer

Buffer	D_b $\mu m\, s^{-1}$	k_+ $\mu M^{-1} s^{-1}$	k_- s^{-1}	K μM
BAPTA	95	600	100	0.17
EGTA	113	1.5	0.3	0.2
Endog	15	50	500	10

Despite the complexity of (74) it retains the advantage of being a single equation. However, if the buffer kinetics are not fast relative to the Ca^{2+} kinetics, the only way to proceed is with numerical simulations of the complete system, a procedure followed by a number of groups (Backx et al., 1989; Sala and Hernández-Cruz, 1990; Nowycky and Pinter, 1993; Falcke, 2003).

6 Calcium Puffs and Stochastic Modeling

In all the models discussed above we have assumed that the release of Ca^{2+} can be modeled as deterministic. However, it is now well known that this is not always the case. In fact, each Ca^{2+} oscillation or wave is built up from a number of stochastic elementary release events, called *puffs*, each of which corresponds to Ca^{2+} release from a single, or a small group of, IPR. In *Xenopus* oocytes, for instance, IPR are arranged in clusters with a density of about 1 per 30 μm^2, with each cluster containing about 25 IPR. At low concentrations of IP_3, punctate release from single clusters occurs while at high $[IP_3]$ these local release events are coordinated into global intracellular waves (Yao et al., 1995; Parker et al., 1996; Parker and Yao, 1996). Detailed studies of puffs have been done principally in *Xenopus* oocytes (Marchant et al., 1999; Sun et al., 1998; Callamaras et al., 1998; Marchant and Parker, 2001) and HeLa cells (Thomas et al., 2000; Bootman et al., 1997a,b). Some typical experimental results are shown in Fig. 16.

These experimental results raise two important modeling questions. Firstly, how should one best model stochastic Ca^{2+} release through a small number of IPR, and, secondly, how can one model the coordination of local release into global events such as intracellular waves? Although such questions are most obvious in the experimental work in oocytes and HeLa cells, they are also important for the study of Ca^{2+} oscillations in other cell types. For instance, Dupont et al. (2000) have estimated that, during a global Ca^{2+} spike in hepatocytes, only about 100 IPR are open at any one time. This means that stochastic effects cannot be ignored even during reponses that appear initially to be wholly deterministic in nature. As Falcke (2004) has pointed out, stochastic effects now appear to be so fundamental and widespread that they raise questions about the applicability of deterministic approaches in general.

6.1 Stochastic IPR Models

The basic assumption behind the modeling of Ca^{2+} puffs is that each IPR can be modeled as a stochastic Markov model while the diffusion of Ca^{2+}, and the other Ca^{2+} fluxes can be modeled deterministically. Given a Markov state model of the IPR (for instance, as shown in Fig. 8 or Fig. 12) we can simulate the model stochastically by choosing a random number at each time step and using that random number to determine the change of state for that time step. The probability for changing state is given by the transition rate multiplied by the time step. Thus, for instance, if we had the transition

$$A \underset{k_{-1}}{\overset{k_1}{\rightleftharpoons}} B \tag{81}$$

then the probability of moving from state A to state B, in time step Δt, is just $k_1 \Delta t$. In general, each transition rate will be a function of $[Ca^{2+}]$ (c) and

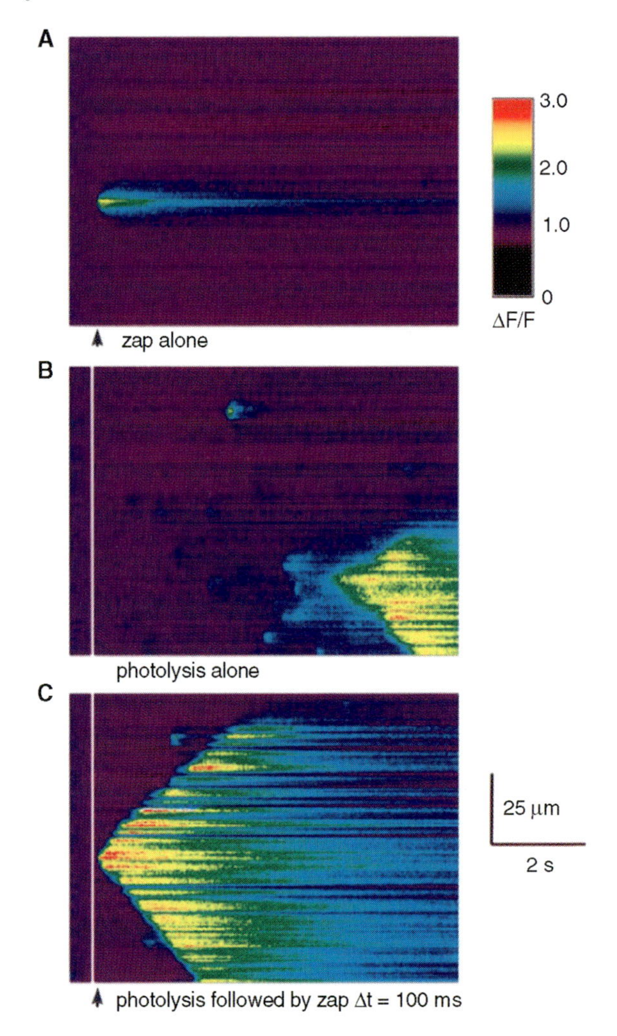

Fig. 16. Calcium waves and puffs caused by release of Ca^{2+} and IP_3. (**A**) release of Ca^{2+} by a UV laser zap results in a single localised Ca^{2+} response. (**B**) photolysis of IP_3 across the entire region causes several Ca^{2+} puffs (see, for instance, the puff at *top center*), followed by an abortive wave about 5 seconds after the release of IP_3. (**C**) photolysis of IP_3 followed by release of Ca^{2+} by a UV laser zap. A global wave is initiated immediately by the additional Ca^{2+} release (Reproduced from Marchant et al., 1999, Fig. 3.)

$[IP_3]$ (p) and thus the transition probabilities are continually changing as the concentrations change. It follows that c and p must be updated at each time step. Such updates are done deterministically, by solving the reaction-diffusion equations for c and p. One thus gets a coupled stochastic/deterministic system of equations to be solved at each time step. When the simulated IPR model is in the open state, the reaction-diffusion equation for $[Ca^{2+}]$ (c) is driven by an additional flux through the IPR. When the simulated IPR model is in some other state, this flux is absent. Thus, effectively, we obtain a reaction-diffusion equation for c that is driven by a stochastically varying flux input.

If the receptor model has a single open state, as in Fig. 8 where the only open state is S_{110}, it is simple to calculate the receptor flux. Whenever the receptor moves into the open state it allows a Ca^{2+} current of given magnitude. However, it is not so simple when the receptor model has multiple open states as in Fig. 12 where both states O and A allow Ca^{2+} flux but state O allows less than state A. There are two possibilities;

1. States O and A both allow Ca^{2+} flux, but one just allows more than the other. In this case one would expect to observe different conductance levels in experiments, which does not appear to be the case.
2. States O and A both flick to and from an open configuration (i.e., each receptor state has both a closed and an open configuration) but state O spends less time in the open configuration than does state A. This would explain how the receptor appears to pass less Ca^{2+} flux when in some states than in others, but no distinct conductance levels are clearly observed.

If the open and shut configurations of each receptor state are assumed to be in instantaneous equilibrium there is no effective difference between these two approaches, at least from the point of view of modeling at this level of detail.

The first stochastic model of an IPR was due to Swillens et al. (1998). Their model is an 18-state model in which the IPR can have 0 or 1 IP_3 bound, 0,1 or 2 activating Ca^{2+} bound, and 0,1 or 2 inactivating Ca^{2+} bound. The receptor states are labeled S_{ijk} where $i = 0, 1$ denotes the occupancy of the IP_3 binding site, $j = 0, 1, 2$ denotes the occupancy of the activating Ca^{2+} binding site, and $k = 0, 1, 2$ denotes the occupancy of the inactivating Ca^{2+} binding site. The receptor is open only when it is in the state S_{120}, i.e., when it has one IP_3 bound, two Ca^{2+} bound to activating sites, and no Ca^{2+} bound to inactivating sites. The steady-state open probability of the model is constrained to fit the experimental data of Bezprozvanny et al. (1991), and it is assumed that Ca^{2+} just diffuses radially away from the mouth of the channel. Calcium can build up to high concentrations at the mouth of the channel, and these local concentrations are used in the stochastic simulation of the IPR model.

Simulations of this model show two things in particular. Firstly, channel openings occur in bursts, as Ca^{2+} diffuses away from the mouth of the channel slowly enough to allow rebinding to an activating site. Secondly, by comparing to experimentally observed distributions of puff amplitudes, Swillens et al., 1999 showed in a later paper that a typical cluster contains approximately 25 receptors, and that within the cluster the IPR are probably separated by no more than 12 nm.

Because of the intensive computations involved in direct stochastic simulation of gating schemes for the IPR, Shuai and Jung reduced a stochastic version of the 8-state model (Sect. 3.5.1) to an equivalent Langevin equation (Shuai and Jung, 2002a,b, 2003). Although 25 receptors in each cluster makes a Langevin equation approach less accurate than direct stochastic simulation, the qualitative behavior agrees well with direct simulation and the errors are within about 10%. Thus, although not many other authors have followed this approach, the method hold promise for fast, reasonably accurate, stochastic simulation of IPR clusters.

6.2 Stochastic Models of Calcium Waves

Of all the results from stochastic modeling, the most intriguing are those of Falcke (2003b) who studied in detail the transition from puffs to waves in *Xenopus* oocytes as $[IP_3]$ is increased. At low $[IP_3]$ only puffs are observed; there is not enough Ca^{2+} release from each cluster to stimulate Ca^{2+} release from neighbouring clusters, and thus the responses are purely local. However, as $[IP_3]$ increases, both the sensitivity of, and the amount of Ca^{2+} released from each IPR increases. This allows for the development of global waves that emerge from a *nucleation* site. However, until $[IP_3]$ gets considerably larger these global events are rare, and in many cases form only abortive waves that progress only a short distance before dying out. The interwave time interval, T_{av}, and its standard deviation ΔT_{av}, is easily determined and decrease as $[IP_3]$ increases. Finally, at high $[IP_3]$ global waves occur regularly with a well-defined period. An example of such a wave, one that has formed a spiral, is shown in Fig. 3A.

Falcke (2003b) showed that all these behaviors can be reproduced by a stochastic version of the 8-state model of Sect. 3.5.1. The long time interval between successive waves when $[IP_3]$ is low is a result almost entirely of the stochastic dynamics. In each time interval there is a chance that one IPR will fire, stimulate the firing of the entire cluster, and thus initiate a global wave, but because of the intercluster separation and the low sensitivity of each IPR such events are rare. Hence in this regime both T_{av} and ΔT_{av} are large. Conversely, when $[IP_3]$ is large, each IPR is much more sensitive to Ca^{2+}, and the flux through each receptor is larger. Thus, firing of a single IPR is nearly always sufficient to stimulate a global wave. In this case, the interwave period is set, not by stochastic effects, but by the intrinsic dynamics of the IPR, i.e., the time taken for the receptor to reactivate and be ready to propagate another wave. These results are most easily seen in Fig. 17 which shows experimental data from Marchant and Parker (2001) and the corresponding simulations of Falcke (2003). The ratio of $\Delta T_{av}/T_{av}$ remains approximately constant over a wide range of T_{av}.

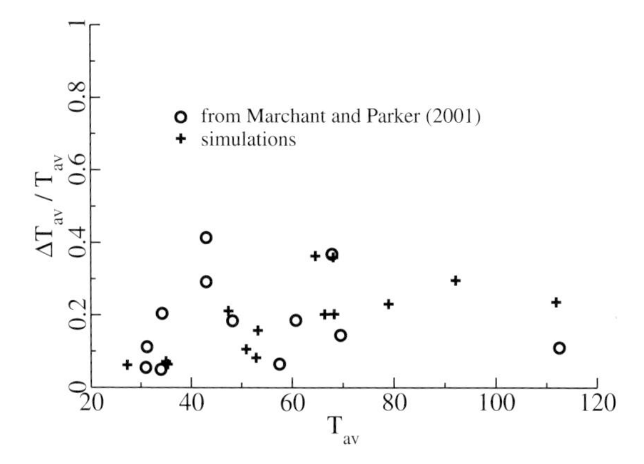

Fig. 17. Plot of the relative standard deviation $\Delta T_{av}/T_{av}$. The experimental data (*open circles*) are from Marchant and Parker (2001), while the simulations are from Falcke (2003b)

What is particularly interesting about these results is that in the simulations the periodic global waves occur for concentrations of IP_3 for which the deterministic version of the model is non-oscillatory. In other words, oscillatory waves are not necessarily the result of oscillatory kinetics. They can result merely from a stochastic process which, every so often, causes a cluster to fire sufficiently strongly that it initiates a global wave. If the standard deviation of the average time between firings is sufficiently small the resulting almost-periodic responses can appear to be the result of an underlying limit cycle, even when no such limit cycle exists. Such results call into serious question the relevance of deterministic approaches to Ca^{2+} waves and oscillations that are caused by small numbers of stochastic IPR. However, the implications have yet to be fully digested.

7 Intercellular Calcium Waves

Not only do Ca^{2+} waves travel across individual cells, they also travel from cell to cell to form intercellular waves that can travel across many cells. One of the earliest examples of such an intercellular wave was discovered by Sanderson et al. (1990); Sanderson et al. (1994) who discovered that, in epithelial cell cultures, a mechanical stimulus (for instance, poking a single cell with a micropipette) can initiate a wave of increased intracellular Ca^{2+} that spreads from cell to cell to form an intercellular wave. Typical experimental results from airway epithelial cells are shown in Fig. 4. The epithelial cell culture forms a thin layer of cells, connected by gap junctions. When a cell in the middle of the culture is mechanically stimulated, the Ca^{2+} in the stimulated cell increases quickly. After a time delay of a second or so, the neighbors of the stimulated cell also show an increase in Ca^{2+}, and this increase spreads sequentially through the culture. An intracellular wave moves across each cell, is delayed at the cell boundary, and then initiates an intracellular wave in the neighboring cell. The intercellular wave moves via the sequential propagation of intracellular waves. Of particular interest here is the fact that in the absence of extracellular Ca^{2+}, the stimulated cell shows no response, but an intercellular wave still spreads to other cells in the culture. It thus appears that a rise in Ca^{2+} in the stimulated cell is not necessary for wave propagation. Neither is a rise in Ca^{2+} sufficient to initiate an intercellular wave. For example, epithelial cells in culture sometimes exhibit spontaneous intracellular Ca^{2+} oscillations, and these oscillations do not spread from cell to cell. Nevertheless, a mechanically stimulated intercellular wave does spread through cells that are spontaneously oscillating.

Intercellular Ca^{2+} waves in glial cultures were also studied by Charles et al. (1991) as well as by Cornell-Bell et al. (1990). Over the past few years there is an increasing body of evidence to show that such intercellular communication between glia, or between glia and neurons, is playing an important role in information processing in the brain (Nedergaard, 1994; Charles, 1998; Vesce et al., 1999; Lin and Bergles, 2004).

Just as there is wide variety of intercellular Ca^{2+} waves in different cell types, so there is a corresponding variety in their mechanism of propagation. Nevertheless, two basic mechanisms are predominant; propagation by the diffusion of an extracellular messenger, or propagation by the diffusion of an intracellular messenger through gap junctions. Sometimes both mechanims operate in combination to drive an intercellular wave (see, for example, Young and Hession, 1997). Most commonly the intracellular messenger is IP_3 or Ca^{2+} (or both), but a much larger array of extracellular messengers, including ATP, ADP and nitric oxide, has been implicated.

There have been few models of intercellular Ca^{2+} waves. The earliest were due to Sneyd et al. (1994, 1995a) and studied the mechanisms underlying mechanically-induced waves. More recent versions of this basic model have been used to study intercellular coupling in hepatocytes (Höfer, 1999; Höfer et al., 2001, 2002; Dupont et al., 2000) and pancreatic acinar cells, while a different approach was taken by Jung et al. (1998). However, detailed studies of the interactions between intracellular and extracellular messengers remain to be done.

References

Allbritton, N. L., Meyer, T., Stryer, L. 1992. Range of messenger action of calcium ion and inositol 1,4,5-trisphosphate. Science 258, 1812–1815.

Atri, A., Amundson, J., Clapham, D., Sneyd, J. 1993. A single-pool model for intracellular calcium oscillations and waves in the *Xenopus laevis* oocyte. Biophys J 65, 1727–1739.

Backx, P. H., de Tombe, P. P., Van Deen, J. H. K., Mulder, B. J. M., ter Keurs, H. E. D. J. 1989. A model of propagating calcium-induced calcium release mediated by calcium diffusion. Journal of General Physiology 93, 963–977.

Berridge, M. J., Galione, A. 1988. Cytosolic calcium oscillators. FASEB Journal 2, 3074–3082.

Berridge, M. J. 1997. Elementary and global aspects of calcium signalling. J Physiol 499 (Pt 2), 291–306.

Berridge, M. J., Bootman, M. D., Roderick, H. L. 2003. Calcium signalling: dynamics, homeostasis and remodelling. Nat Rev Mol Cell Biol 4, 517–529.

Bezprozvanny, I., Watras, J., Ehrlich, B. E. 1991. Bell-shaped calcium-response curves of Ins(1,4,5)P_3- and calcium-gated channels from endoplasmic reticulum of cerebellum. Nature 351, 751–754.

Boitano, S., Dirksen, E. R., Sanderson, M. J. 1992. Intercellular propagation of calcium waves mediated by inositol trisphosphate. Science 258, 292–294.

Bootman, M., Niggli, E., Berridge, M., Lipp, P. 1997a. Imaging the hierarchical Ca^{2+} signalling system in HeLa cells. J Physiol 499 (Pt 2), 307–314.

Bootman, M. D., Berridge, M. J., Lipp, P. 1997b. Cooking with calcium: the recipes for composing global signals from elementary events. Cell 91, 367–373.

Bugrim, A., Fontanilla, R., Eutenier, B. B., Keizer, J., Nuccitelli, R. 2003. Sperm initiate a Ca^{2+} wave in frog eggs that is more similar to Ca^{2+} waves initiated by IP_3 than by Ca^{2+}. Biophys J 84, 1580–1590.

Callamaras, N., Marchant, J. S., Sun, X. P., Parker, I. 1998. Activation and coordination of InsP_3-mediated elementary Ca^{2+} events during global Ca^{2+} signals in Xenopus oocytes. J Physiol 509 (Pt 1), 81–91.

Carafoli, E., Santella, L., Branca, D., Brini, M. 2001. Generation, control, and processing of cellular calcium signals. Crit Rev Biochem Mol Biol 36, 107–260.

Carafoli, E. 2002. Calcium signaling: a tale for all seasons. Proc Natl Acad Sci U S A 99, 1115–1122.

Charles, A. C., Merrill, J. E., Dirksen, E. R., Sanderson, M. J. 1991. Intercellular signaling in glial cells: calcium waves and oscillations in response to mechanical stimulation and glutamate. Neuron 6, 983–992.

Charles, A. C., Naus, C. C. G., Zhu, D., Kidder, G. M., Dirksen, E. R., Sanderson, M. J. 1992. Intercellular calcium signaling via gap junctions in glioma cells. Journal of Cell Biology 118, 195–201.

Charles, A. 1998. Intercellular calcium waves in glia. Glia 24, 39–49.

Cheer, A., Nuccitelli, R., Oster, G. F., Vincent, J.-P. 1987. Cortical waves in vertebrate eggs I: the activation waves. J Theor Biol 124, 377–404.

Cheng, H., Lederer, W. J., Cannell, M. B. 1993. Calcium sparks: elementary events underlying excitation-contraction coupling in heart muscle. Science 262, 740–744.

Cheng, H., Lederer, M. R., Lederer, W. J., Cannell, M. B. 1996. Calcium sparks and $[Ca^{2+}]_i$ waves in cardiac myocytes. Am J Physiol 270, C148–159.

Colegrove, S. L., Albrecht, M. A., Friel, D. D. 2000. Quantitative analysis of mitochondrial Ca^{2+} uptake and release pathways in sympathetic neurons. Reconstruction of the recovery after depolarization-evoked $[Ca^{2+}]_i$ elevations. J Gen Physiol 115, 371–388.

Cornell-Bell, A. H., Finkbeiner, S. M., Cooper, M. S., Smith, S. J. 1990. Glutamate induces calcium waves in cultured astrocytes: long-range glial signaling. Science 247, 470–473.

Cuthbertson, K. S. R., Chay, T. R. 1991. Modelling receptor-controlled intracellular calcium oscillators. Cell Calcium 12, 97–109.

De Young, G. W., Keizer, J. 1992. A single pool IP_3-receptor based model for agonist stimulated Ca^{2+} oscillations. Proc Natl Acad Sci U S A 89, 9895–9899.

Demer, L. L., Wortham, C. M., Dirksen, E. R., Sanderson, M. J. 1993. Mechanical stimulation induces intercellular calcium signalling in bovine aortic endothelial cells. Am. J. Physiol. (Heart & Circ. Physiol.) 33, M2094–M2102.

Dolmetsch, R. E., Xu, K., Lewis, R. S. 1998. Calcium oscillations increase the efficiency and specificity of gene expression. Nature 392, 933–936.

Duffy, M. R., Britton, N. F., Murray, J. D. 1980. Spiral wave solutions of practical reaction-diffusion systems. SIAM Journal on Applied Mathematics 39, 8–13.

Dufour, J.-F., Arias, I. M., Turner, T. J. 1997. Inositol 1,4,5-trisphosphate and calcium regulate the calcium channel function of the hepatic inositol 1,4,5-trisphosphate receptor. J Biol Chem 272, 2675–2681.

Dupont, G., Erneux, C. 1997. Simulations of the effects of inositol 1,4,5-trisphosphate 3-kinase and 5-phosphatase activities on Ca^{2+} oscillations. Cell Calcium 22, 321–331.

Dupont, G., Tordjmann, T., Clair, C., Swillens, S., Claret, M., Combettes, L. 2000. Mechanism of receptor-oriented intercellular calcium wave propagation in hepatocytes. Faseb J 14, 279–289.

Dupont, G., Koukoui, O., Clair, C., Erneux, C., Swillens, S., Combettes, L. 2003. Ca^{2+} oscillations in hepatocytes do not require the modulation of $InsP_3$ 3-kinase activity by Ca^{2+}. FEBS Lett 534, 101–105.

Falcke, M., Hudson, J. L., Camacho, P., Lechleiter, J. D. 1999. Impact of mitochondrial Ca^{2+} cycling on pattern formation and stability. Biophys J 77, 37–44.

Falcke, M., Or-Guil, M., Bar, M. 2000. Dispersion gap and localized spiral waves in a model for intracellular Ca^{2+} dynamics. Phys. Rev. Lett. 84, 4753–4756.

Falcke, M. 2003. Buffers and oscillations in intracellular Ca^{2+} dynamics. Biophys J 84, 28–41.

Falcke, M. 2003b. On the role of stochastic channel behavior in intracellular Ca^{2+} dynamics. Biophys J 84, 42–56.

Falcke, M. 2004. Reading the patterns in living cells – the physics of Ca^{2+} signaling. Advances in Physics 53, 255–440.

Favre, C. J., Schrenzel, J., Jacquet, J., Lew, D. P., Krause, K. H. 1996. Highly supralinear feedback inhibition of Ca^{2+} uptake by the Ca^{2+} load of intracellular stores. J. Biol. Chem. 271, 14925–14930.

Finch, E. A., Turner, T. J., Goldin, S. M. 1991. Calcium as a coagonist of inositol 1,4,5-trisphosphate-induced calcium release. Science 252, 443–446.

Fontanilla, R. A., Nuccitelli, R. 1998. Characterization of the sperm-induced calcium wave in Xenopus eggs using confocal microscopy. Biophys J. 75, 2079–2087.

Friel, D. D. 2000. Mitochondria as regulators of stimulus-evoked calcium signals in neurons. Cell Calcium 28, 307–316.

Gilkey, J. C., Jaffe, L. F., Ridgway, E. B., Reynolds, G. T. 1978. A free calcium wave traverses the activating egg of the medaka *oryzias latipes*. J Cell Biol 76, 448–466.

Grubelnik, V., Larsen, A. Z., Kummer, U., Olsen, L. F., Marhl, M. 2001. Mitochondria regulate the amplitude of simple and complex calcium oscillations. Biophys Chem 94, 59–74.

Hajnóczky, G., Thomas, A. P. 1997. Minimal requirements for calcium oscillations driven by the IP_3 receptor. Embo J 16, 3533–3543.

Hirose, K., Kadowaki, S., Tanabe, M., Takeshima, H., Iino, M. 1999. Spatiotemporal dynamics of inositol 1,4,5-trisphosphate that underlies complex Ca^{2+} mobilization patterns. Science 284, 1527–1530.

Hofer, A. M., Brown, E. M. (2003). Extracellular calcium sensing and signalling, Nature Reviews (Molecular Cell Biology), 4, 530–538.

Houart, G., Dupont, G., Goldbeter, A. 1999. Bursting, chaos and birhythmicity originating from self-modulation of the inositol 1,4,5-trisphosphate signal in a model for intracellular Ca^{2+} oscillations. Bull. Math. Biol. 61, 507–530.

Höfer, T. 1999. Model of intercellular calcium oscillations in hepatocytes: synchronization of heterogeneous cells. Biophys J 77, 1244–1256.

Höfer, T., Politi, A., Heinrich, R. 2001. Intercellular Ca^{2+} wave propagation through gap-junctional Ca^{2+} diffusion: a theoretical study. Biophys J 80, 75–87.

Höfer, T., Venance, L., Giaume, C. 2002. Control and plasticity of intercellular calcium waves in astrocytes: a modeling approach. J Neurosci 22, 4850–4859.

Irving, M., Maylie, J., Sizto, N. L., Chandler, W. K. 1990. Intracellular diffusion in the presence of mobile buffers: application to proton movement in muscle. Biophysical Journal 57, 717–721.

Izu, L. T., Wier, W. G., Balke, C. W. 2001. Evolution of cardiac calcium waves from stochastic calcium sparks. Biophys J 80, 103–120.

Jung, P., Cornell-Bell, A., Madden, K. S., Moss, F. 1998. Noise-induced spiral waves in astrocyte syncytia show evidence of self- organized criticality. J Neurophysiol 79, 1098–1101.

Keener, J. P., Sneyd, J. (1998) *Mathematical Physiology* (Springer-Verlag, New York)

Keizer, J., De Young, G. W. 1994. Simplification of a realistic model of IP_3-induced Ca^{2+} oscillations. J theor Biol 166, 431–442.

Kidd, J. F., Fogarty, K. E., Tuft, R. A., Thorn, P. 1999. The role of Ca^{2+} feedback in shaping $InsP_3$-evoked Ca^{2+} signals in mouse pancreatic acinar cells. J Physiol 520 Pt 1, 187–201.

Kopell, N., Howard, L. N. 1973. Plane wave solutions to reaction-diffusion equations. Studies in Applied Mathematics 52, 291–328.

Lane, D. C., Murray, J. D., Manoranjan, V. S. 1987. Analysis of wave phenomena in a morphogenetic mechanochemical model and an application to post-fertilisation waves on eggs. IMA J Math Med Biol 4, 309–331.

Lechleiter, J., Girard, S., Clapham, D., Peralta, E. 1991a. Subcellular patterns of calcium release determined by G protein-specific residues of muscarinic receptors. Nature 350, 505–508.

Lechleiter, J., Girard, S., Peralta, E., Clapham, D. 1991b. Spiral calcium wave propagation and annihilation in *Xenopus laevis* oocytes. Science 252, 123–126.

Lechleiter, J., Clapham, D. 1992. Molecular mechanisms of intracellular calcium excitability in X. laevis oocytes. Cell 69, 283–294.

Li, Y.-X., Stojilković, S. S., Keizer, J., Rinzel, J. 1997. Sensing and refilling calcium stores in an excitable cell. Biophys J 72, 1080–1091.

Li, W., Llopis, J., Whitney, M., Zlokarnik, G., Tsien, R. Y. 1998. Cell-permeant caged $InsP_3$ ester shows that Ca^{2+} spike frequency can optimize gene expression. Nature 392, 936–941.

Lin, S. C., Bergles, D. E. 2004. Synaptic signaling between neurons and glia. Glia 47, 290–298.

Lytton, J., Westlin, M., Burk, S. E., Shull, G. E., MacLennan, D. H. 1992. Functional comparisons between isoforms of the sarcoplasmic or endoplasmic reticulum family of calcium pumps. J. Biol. Chem. 267, 14483–14489.

MacLennan, D. H., Rice, W. J., Green, N. M. 1997. The mechanism of Ca^{2+} transport by sarco(endo)plasmic reticulum Ca^{2+}-ATPases. J Biol Chem 272, 28815–28818.

Maginu, K. 1985. Geometrical characteristics associated with stability and bifurcations of periodic travelling waves in reaction-diffusion equations. SIAM Journal on Applied Mathematics 45, 750–774.

Marchant, J., Callamaras, N., Parker, I. 1999. Initiation of IP_3-mediated Ca^{2+} waves in Xenopus oocytes. Embo J 18, 5285–5299.

Marchant, J. S., Parker, I. 2001. Role of elementary Ca^{2+} puffs in generating repetitive Ca^{2+} oscillations. Embo J 20, 65–76.

Marhl, M., Haberichter, T., Brumen, M., Heinrich, R. 2000. Complex calcium oscillations and the role of mitochondria and cytosolic proteins. Biosystems 57, 75–86.

McKenzie, A., Sneyd, J. 1998. On the formation and breakup of spiral waves of calcium. Int. J. Bif. Chaos 8, 2003–2012.

Meyer, T., Stryer, L. 1988. Molecular model for receptor-stimulated calcium spiking. Proc. Natl. Acad. Sci. USA 85, 5051–5055.

Murray, J. D. (1989) *Mathematical Biology* (Springer Verlag, Berlin, Heidelberg, New York)

Naraghi, M., Neher, E. 1997. Linearized buffered Ca^{2+} diffusion in microdomains and its implications for calculation of $[Ca^{2+}]$ at the mouth of a calcium channel. J Neurosci 17, 6961–6973.

Nash, M. S., Young, K. W., Challiss, R. A., Nahorski, S. R. 2001. Intracellular signalling. Receptor-specific messenger oscillations. Nature 413, 381–382.

Nathanson, M. H., Burgstahler, A. D., Mennone, A., Fallon, M. B., Gonzalez, C. B., Saez, J. C. 1995. Ca^{2+} waves are organized among hepatocytes in the intact organ. Am. J. Physiol. 269, G167–G171.

Nedergaard, M. 1994. Direct signaling from astrocytes to neurons in cultures of mammalian brain cells. Science 263, 1768–1771.

Neu, J. C. 1979. Chemical waves and the diffusive coupling of limit cycle oscillators. SIAM J. Appl. Math. 36, 509–515.

Nowycky, M. C., Pinter, M. J. 1993. Time courses of calcium and calcium-bound buffers following calcium influx in a model cell. Biophysical Journal 64, 77–91.

Nuccitelli, R., Yim, D. L., Smart, T. 1993. The sperm-induced Ca^{2+} wave following fertilization of the Xenopus egg requires the production of Ins(1,4,5)P_3. Dev Biol 158, 200–212.

Orrenius, S., Zhivotovsky, B., Nicotera, P. (2003). Regulation of cell death: the calcium-apoptosis link, Nature Reviews (Molecular Cell Biology), 4, 552–565.

Parker, I., Ivorra, I. 1990. Inhibition by Ca^{2+} of inositol trisphosphate-mediated Ca^{2+} liberation: a possible mechanism for oscillatory release of Ca^{2+}. Proc Natl Acad Sci U S A 87, 260–264.

Parker, I., Choi, J., Yao, Y. 1996. Elementary events of $InsP_3$-induced Ca^{2+} liberation in Xenopus oocytes: hot spots, puffs and blips. Cell Calcium 20, 105–121.

Parker, I., Yao, Y. 1996. Ca^{2+} transients associated with openings of inositol trisphosphate-gated channels in Xenopus oocytes. J Physiol 491 (Pt 3), 663–668.

Parys, J. B., Sernett, S. W., DeLisle, S., Snyder, P. M., Welsh, M. J., Campbell, K. P. 1992. Isolation, characterization, and localization of the inositol 1,4,5-

trisphosphate receptor protein in *Xenopus laevis* oocytes. J Biol Chem 267, 18776–18782.

Ridgway, E. B., Gilkey, J. C., Jaffe, L. F. 1977. Free calcium increases explosively in activating medaka eggs. Proc Natl Acad Sci U S A 74, 623–627.

Rinzel, J. (1985) in *Ordinary and partial differential equations*, eds. Sleeman, B.D. and Jarvis, R.J. (Springer-Verlag, New York)

Robb-Gaspers, L. D., Thomas, A. P. 1995. Coordination of Ca^{2+} signaling by intercellular propagation of Ca^{2+} waves in the intact liver. J Biol Chem 270, 8102–8107.

Rooney, T. A., Thomas, A. P. 1993. Intracellular calcium waves generated by Ins(1,4,5)P$_3$-dependent mechanisms. Cell Calcium 14, 674–690.

Røttingen, J., Iversen, J. G. 2000. Ruled by waves? Intracellular and intercellular calcium signalling. Acta Physiol Scand 169, 203–219.

Sala, F., Hernández-Cruz, A. 1990. Calcium diffusion modeling in a spherical neuron: relevance of buffering properties. Biophysical Journal 57, 313–324.

Sanderson, M. J., Charles, A. C., Dirksen, E. R. 1990. Mechanical stimulation and intercellular communication increases intracellular Ca^{2+} in epithelial cells. Cell Reg 1, 585–596.

Sanderson, M. J., Charles, A. C., Boitano, S., Dirksen, E. R. 1994. Mechanisms and function of intercellular calcium signaling. Molecular and Cellular Endocrinology 98, 173–187.

Schuster, S., Marhl, M., Höfer, T. 2002. Modelling of simple and complex calcium oscillations. From single-cell responses to intercellular signalling. Eur J Biochem 269, 1333–1355.

Selivanov, V. A., Ichas, F., Holmuhamedov, E. L., Jouaville, L. S., Evtodienko, Y. V., Mazat, J. P. 1998. A model of mitochondrial Ca^{2+}-induced Ca^{2+} release simulating the Ca^{2+} oscillations and spikes generated by mitochondria. Biophys Chem 72, 111–121.

Shen, P., Larter, R. 1995. Chaos in intracellular Ca^{2+} oscillations in a new model for non-excitable cells. Cell Calcium 17, 225–232.

Shuai, J. W., Jung, P. 2002a. Optimal intracellular calcium signaling. Phys Rev Lett 88, 068102.

Shuai, J. W., Jung, P. 2002b. Stochastic properties of Ca^{2+} release of inositol 1,4,5-trisphosphate receptor clusters. Biophys J 83, 87–97.

Shuai, J. W., Jung, P. 2003. Optimal ion channel clustering for intracellular calcium signaling. Proc Natl Acad Sci U S A 100, 506–510.

Shuttleworth, T. J. 1999. What drives calcium entry during $[Ca^{2+}]_i$ oscillations? – challenging the capacitative model. Cell Calcium 25, 237–246.

Smith, G. D., Wagner, J., Keizer, J. 1996. Validity of the rapid buffering approximation near a point source of calcium ions. Biophys J 70, 2527–2539.

Smith, G. D. 1996. Analytical steady-state solution to the rapid buffering approximation near an open Ca^{2+} channel. Biophys J 71, 3064–3072.

Smith, G. D., Keizer, J. E., Stern, M. D., Lederer, W. J., Cheng, H. 1998. A simple numerical model of calcium spark formation and detection in cardiac myocytes. Biophys J 75, 15–32.

Smith, G. D., Dai, L., Miura, R. M., Sherman, A. 2001. Asymptotic analysis of buffered calcium diffusion near a point source. SIAM J on Appl Math 61, 1816–1838.

Sneyd, J., Girard, S., Clapham, D. 1993. Calcium wave propagation by calcium-induced calcium release: an unusual excitable system. Bulletin of Mathematical Biology 55, 315–344.
Sneyd, J., Charles, A. C., Sanderson, M. J. 1994. A model for the propagation of intercellular calcium waves. Am J Physiol (Cell Physiol) 266, C293–C302.
Sneyd, J., Wetton, B., Charles, A. C., Sanderson, M. J. 1995a. Intercellular calcium waves mediated by diffusion of inositol trisphosphate: a two-dimensional model. American Journal of Physiology (Cell Physiology) 268, C1537–C1545.
Sneyd, J., Keizer, J., Sanderson, M. J. 1995b. Mechanisms of calcium oscillations and waves: a quantitative analysis. FASEB J 9, 1463–1472.
Sneyd, J., Dufour, J. F. 2002. A dynamic model of the type-2 inositol trisphosphate receptor. Proc Natl Acad Sci U S A 99, 2398–2403.
Sneyd, J., Tsaneva-Atanasova, K., Yule, D. I., Thompson, J. L., Shuttleworth, T. J. 2004a. Control of calcium oscillations by membrane fluxes. Proc Natl Acad Sci U S A 101, 1392–1396.
Sneyd, J., Falcke, M. 2004. Models of the inositol trisphosphate receptor. Prog Biophys Mol Biol, in press.
Sneyd, J., Falcke, M., Dufour, J. F., Fox, C. 2004b. A comparison of three models of the inositol trisphosphate receptor. Prog Biophys Mol Biol 85, 121–140.
Sobie, E. A., Dilly, K. W., dos Santos Cruz, J., Lederer, W. J., Jafri, M. S. 2002. Termination of cardiac Ca^{2+} sparks: an investigative mathematical model of calcium-induced calcium release. Biophys J 83, 59–78.
Soeller, C., Cannell, M. B. 1997. Numerical simulation of local calcium movements during L-type calcium channel gating in the cardiac diad. Biophys J 73, 97–111.
Soeller, C., Cannell, M. B. 2002. Estimation of the sarcoplasmic reticulum Ca^{2+} release flux underlying Ca^{2+} sparks. Biophys J 82, 2396–2414.
Stern, M. D. 1992. Buffering of calcium in the vicinity of a channel pore. Cell Calcium 13, 183–192.
Sun, X. P., Callamaras, N., Marchant, J. S., Parker, I. 1998. A continuum of $InsP_3$-mediated elementary Ca^{2+} signalling events in Xenopus oocytes. J Physiol 509 (Pt 1), 67–80.
Swatton, J. E., Taylor, C. W. 2002. Fast biphasic regulation of type 3 inositol trisphosphate receptors by cytosolic calcium. J Biol Chem 277, 17571–17579.
Swillens, S., Mercan, D. 1990. Computer simulation of a cytosolic calcium oscillator. Biochem. Journal 271, 835–838.
Swillens, S., Champeil, P., Combettes, L., Dupont, G. 1998. Stochastic simulation of a single inositol 1,4,5-trisphosphate-sensitive Ca^{2+} channel reveals repetitive openings during "blip-like" Ca^{2+} transients. Cell Calcium 23, 291–302.
Swillens, S., Dupont, G., Combettes, L., Champeil, P. 1999. From calcium blips to calcium puffs: theoretical analysis of the requirements for interchannel communication. Proc Natl Acad Sci U S A 96, 13750–13755.
Tang, Y., Stephenson, J. L., Othmer, H. G. 1996. Simplification and analysis of models of calcium dynamics based on IP_3-sensitive calcium channel kinetics. Biophys J 70, 246–263.
Taylor, C. W. 1998. Inositol trisphosphate receptors: Ca^{2+}-modulated intracellular Ca^{2+} channels. Biochim Biophys Acta 1436, 19–33.
Taylor, C. W., Laude, A. J. 2002. IP_3 receptors and their regulation by calmodulin and cytosolic Ca^{2+}. Cell Calcium 32, 321–334.

Thomas, A. P., Bird, G. S. J., Hajnóczky, G., Robb-Gaspers, L. D., Putney, J. W. J. 1996. Spatial and temporal aspects of cellular calcium signaling. FASEB J 10, 1505–1517.

Thomas, D., Lipp, P., Tovey, S. C., Berridge, M. J., Li, W., Tsien, R. Y., Bootman, M. D. 2000. Microscopic properties of elementary Ca^{2+} release sites in non-excitable cells. Curr Biol 10, 8–15.

Tordjmann, T., Berthon, B., Claret, M., Combettes, L. 1997. Coordinated intercellular calcium waves induced by noradrenaline in rat hepatocytes: dual control by gap junction permeability and agonist. Embo J 16, 5398–5407.

Tordjmann, T., Berthon, B., Jacquemin, E., Clair, C., Stelly, N., Guillon, G., Claret, M., Combettes, L. 1998. Receptor-oriented intercellular calcium waves evoked by vasopressin in rat hepatocytes. Embo J 17, 4695–4703.

Tse, A., Tse, F. W., Almers, W., Hille, B. 1993. Rhythmic exocytosis stimulated by GnRH-induced calcium oscillations in rat gonadotropes. Science 260, 82–84.

Vesce, S., Bezzi, P., Volterra, A. 1999. The active role of astrocytes in synaptic transmission. Cell Mol Life Sci 56, 991–1000.

Wagner, J., Keizer, J. 1994. Effects of rapid buffers on Ca^{2+} diffusion and Ca^{2+} oscillations. Biophysical Journal 67, 447–456.

Wagner, J., Li, Y.-X., Pearson, J., Keizer, J. 1998. Simulation of the fertilization Ca^{2+} wave in Xenopus laevis eggs. Biophys J. 75, 2088–2097.

Webb, S. E., Miller, A. L. (2003). Calcium signalling during embryonic development, Nature Reviews (Molecular Cell Biology), 4, 539–551.

Yao, Y., Choi, J., Parker, I. 1995. Quantal puffs of intracellular Ca^{2+} evoked by inositol trisphosphate in Xenopus oocytes. J Physiol 482 (Pt 3), 533–553.

Young, R. C., Hession, R. O. 1997. Paracrine and intracellular signaling mechanisms of calcium waves in cultured human uterine myocytes. Obstet Gynecol 90, 928–932.

Young, K. W., Nash, M. S., Challiss, R. A., Nahorski, S. R. 2003. Role of Ca^{2+} feedback on single cell inositol 1,4,5-trisphosphate oscillations mediated by G-protein-coupled receptors. J Biol Chem 278, 20753–20760.

Yule, D. I., Stuenkel, E., Williams, J. A. 1996. Intercellular calcium waves in rat pancreatic acini: mechanism of transmission. Am. J. Physiol. 271, C1285–1294.

Integrated Calcium Management in Cardiac Myocytes

T.R. Shannon

Department of Molecular Biophysics and Physiology, Rush University, Chicago, IL

1 Introduction

Cardiac myocyte excitation-contraction coupling (ECC, Table 1) is an intricate process by which many proteins and substances interact to form a complex but well-tuned system. The regulation of this system is essential to modulation of contractile activity.

A great deal of cellular energy within the system is directed toward management of Ca. The rise and fall of [Ca] within the cell is central for initiation of contraction and relaxation, respectively. Thus Ca may be considered to be among the most important second messengers within the ventricular myocyte.

This chapter will describe the ways in which Ca is managed within the cardiac myocyte and how that Ca results in the generation of force. The geometry of the cell is described first to use as a basis for understanding Ca handling in each cellular compartment. A description of action potential generation follows as the stimulus for Ca entry into the myocyte and for its subsequent release from storage and rise within the cytosol. The subsequent decline in [Ca] is described and, finally, a brief description of the effects of Ca upon the myofilaments and the resulting force generation is discussed. It is hoped that the chapter will serve as a general overview for the novice in the area of cardiac ECC from which study in further detail may proceed.

2 Geometry of the Cardiac Myocyte

Though there is a large variation in size, the typical cardiac myocyte is a cylindrical cell approximately 100 µm in length by 10 µm in diameter. The cell volume of approximately 30 pL is surrounded by a cell membrane which in muscle cells is known as the sarcolemma (SL). Large invaginations of this membrane into the cell, known as "T-tubules", approximately double the surface to volume ratio (Fig. 1).

Table 1. Abbreviations used in the text

Abbreviation	Definition
AP	action potential
CICR	Ca-induced Ca release
$[Ca]_i$	Free cytosolic Ca
$[Ca]_{SR}$	Free SR Ca
E_I	Reversal potential for ion, I
ECC	excitation-contraction coupling
E_m	membrane voltage
$E_{Na/Ca}$	Reversal potential for the Na-Ca exchanger
I_{Ca}	L-type Ca current
$I_{Ca,T}$	T-type Ca current
I_{K1}	Inward rectifier K current
I_{Na}	Fast Na current
I_{TO}	Transient outward K current
I_{Kr}	Rapidly activating K Current
LTCC	L-type Ca channel
$[Na]_i$	Free cellular [Na]
PLB	Phospholamban
RyR	Ryanodine receptor
SL	Sarcolemma
SR	Sarcoplasmic reticulum
SSL	Subsarcolemmal compartment
TnC	Troponin C
TnI	Troponin I
TnT	Troponin T
V_{max}	Maximum transport rate

The volume of the cell can be split into at least three physical compartments, the cytoplasm, the sarcoplasmic reticulum (SR) and the mitochondria. The role of mitochondria in the Ca dynamics associated with cardiac ECC is controversial but they have traditionally been considered to be a minor player and won't be discussed in depth in this chapter except as a minor Ca sink. The cytosol, in turn can be split into a minimum of two distinct physiological compartments, the junctional cleft and the bulk cytosol, though more and more evidence is accumulating for the existence of a third compartment, the subsarcolemmal (SSL) compartment (Trafford et al., 1995; Weber et al., 2002, Fig. 1). It is important to emphasize that the SSL is not an anatomical compartment, but a physiological one where the [Ca] tends to be higher due to diffusional gradients (see below).

The SR space occupies approximately 3% of the total myocyte volume (Page et al., 1971). The primary function of this organelle is to store Ca for release upon cellular excitation. Ca within this compartment can be bound to the low affinity Ca binding protein calsequestrin. Calsequestrin is found in the areas of the SR which are specialized for release, the terminal cisternae,

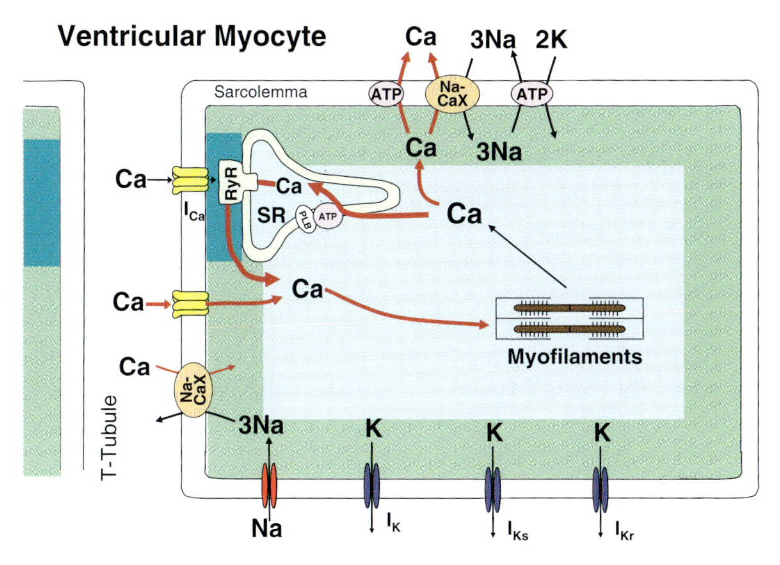

Fig. 1. Diagram of the cell with relevant compartments, channels and transporters. Cardiac ECC takes place via CICR in the junctional cleft (*dark green*). The Ca diffuses into the bulk cytosol (*blue*) where it initiates the process of contraction at the myofilaments. Relaxation takes place as Ca in transported back into the SR or out into the extracellular space

and this is where the majority of the Ca is likely stored between release events. Total Ca concentration within the SR is about 3-5 mM though the concentration within the terminal cisternae may be as high as 90 mM or more. Most of this Ca is bound with only approximately 1 mM free.

The SR lumen is bounded by the SR membrane, which contains proteins which mediate both the release and re-uptake of Ca. The SR release channel, or ryanodine receptor (RyR), is found almost entirely within the part of the SR membrane which forms the boundary of the terminal cisternae and which communicates with the junctional cleft.

The junctional cleft is a very narrow space between the SL and the SR membrane where the terminal cisternae approach very close to the cell membrane. About 11% of the SL membrane and 10% of the SR membrane bound junctional clefts within the cell with about 15 nm of space in between. The total occupied volume is about 0.1% for a space of these dimensions (Page, 1978; Soeller and Cannell, 1997). The size and shape of individual junctions within the cell is not known with certainty but considering them to be cylindrical with a radius of 50–250 nm (and a height of 15 nm) seems to be a reasonable estimate.

During the twitch Ca transient there are spatial gradients near the SL membrane. These gradients are likely made larger by diffusion restrictions which are caused by the mitochondria and the paracrystaline array in the bulk cytosol. Ca-dependent proteins which are located in or near the membrane are therefore likely responding to a [Ca] which is very different from that which is found in the bulk cytosolic compartment.

In order to account for the difference between the Ca of the bulk cytosol and the Ca just underneath the membrane, another physiological compartment within the cell, the subsarcolemmal compartment (SSL) may be defined. The compartment occupies all of the volume under the SL membrane in all areas which are not junctional (i.e. about 89% of the SL membrane area). Because the SSL is a conceptional convenience which is used to account for what is in reality a continuous [Ca] gradient, the depth of the compartment cannot be definitively delineated. A depth of 45 nm (2% of the cell volume) has been used to model the compartment for rabbit ventricular myocytes (Shannon et al., 2004).

Various Ca binding proteins are distributed in each compartment (Table 2). The primary SR Ca buffer is, naturally, calsequestrin. Among the other cellular proteins in the various compartments, those associated with the myofilaments and cellular contraction are the most physiologically important. Troponin C (TnC) binds the most Ca during a typical contractile cycle. Often times when modeling physiological experiments, additional buffers such as Ca-dependent fluorescent dyes (e.g indo-1 and fluo-3) must also be accounted for.

3 Overview of Excitation-Contraction Coupling

A simple working model may be constructed as background to describe the process of ECC as a basis for further discussion. ECC is initiated by depolarization of the SL membrane by the cardiac action potential (AP). The increased voltage results in opening of L-type Ca channels (LTCC) within the SL membrane. These channels are localized such that about 90% are in the junctional SL (Scriven et al., 2000).

The influx of Ca through these channels results in a Ca spike within the junctional cleft and this Ca binds to the RyR located there. The RyR open in response to this binding and Ca efflux from the SR though this channel occurs via a process termed Ca-induced Ca release (CICR). Both Ca released from the SR and Ca from the LTCC influx diffuse from the junctional space and the SSL into the cytoplasm where it binds to troponin C (TnC) and the process of contraction is initiated.

The process of relaxation takes place as Ca is removed from the cytosol. This removal takes place primarily through Ca transporters i.e. the SR Ca pump, the Na-Ca exchanger or the SL Ca pump. At steady-state, the amount of Ca extruded into the extracellular environment by the Na-Ca exchanger and the SL Ca pump is equal to the amount which comes in during excitation through the LTCC. In turn, the same amount is taken back up into the SR as is released.

Table 2. Parameters for Ca buffers

Buffer	Compartment	B_{max} (μmol/l cytosol*)	K_d (μM)	k_{off} (s^{-1})	k_{on} ($\mu M^{-1}s^{-1}$)
troponin C	cytosol	70	0.6	19.6	32.7
troponin C Ca-Mg (Ca)	cytosol	140	0.0135	0.032	2.37
troponin C Ca-Mg (Mg)	cytosol	140	1111	3.33	0.003
calmodulin	cytosol	24	7	238	34
myosin (Ca)	cytosol	140	0.0333	0.46	13.8
myosin (Mg)	cytosol	140	3.64	0.057	0.0157
SR	cytosol	19	0.6	60	100
SL	SL	37.4	13	1300	100
SL	junction	4.6	13	1300	100
SLHigh	SL	13.4	0.3	30	100
SLHigh	junction	1.65	0.3	30	100
Indo-1	cytosol	25	0.6	60	100
Indo-1	SL	0.77	0.6	60	100
Indo-1	junction	0.02	0.6	60	100
Fluo-3	cytosol	25	1.1	110	100
Fluo-3	SL	0.77	1.1	110	100
Fluo-3	junction	0.02	1.1	110	100
calsequestrin	SR	140	650	65000	100

*Concentrations in the actual compartments can be calculated assuming 11% of membrane bound components are in the junctional cleft and 89% are in the SL compartment. Table from Shannon et al. (2004)

4 Action Potentials

As can be seen in Fig. 2, the AP shape varies with tissue and region within the heart. The variation is due to differences in protein expression and cellular function from region to region. Voltage across the SL is always referenced with respect to the interior of the cell. Because the inside of the cell is negative relative to the outside at the beginning of the AP, this voltage, the resting potential, is said to be negative. From the resting potential, the cell rapidly depolarizes, comes to a plateau, then repolarizes back to the resting potential.

During a typical cardiac cycle, the sino-atrial node sets the pace by slowly depolarizing until a threshold potential is reached. From this small area in the groove between the superior vena cava and the right atrium, the AP is propagated outward. Other cell types within the heart also will depolarize at rest but more slowly such that a propagating AP will normally depolarize these cells before they reach threshold.

The AP propagates through the atria and to a small area between the atria and the ventricles known as the atrio-ventricular node. This area is the only portion of electrical communication between these two tissues. From there,

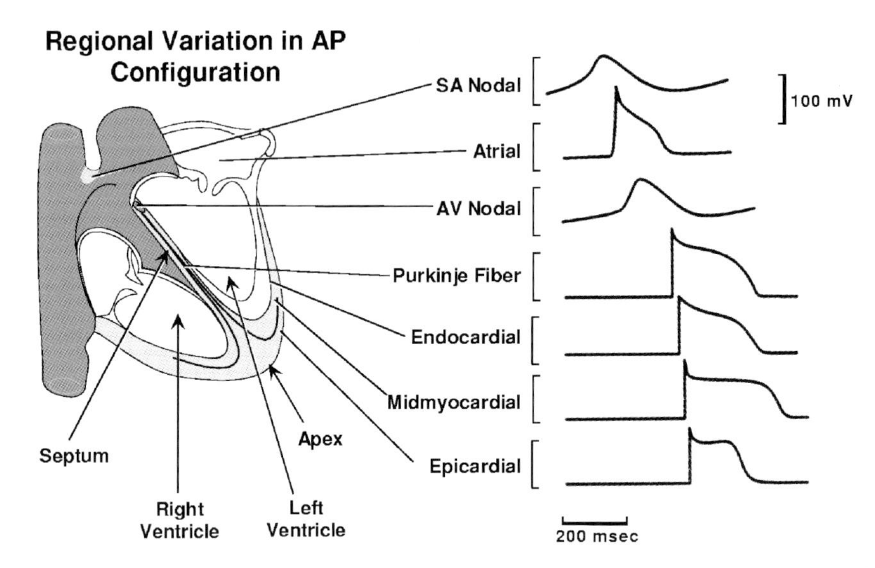

Fig. 2. Action potentials in various portions of the heart. The shape varies with the region and function. The AP within the heart propagates from the sino-atrial node, through the atria to the atrio-ventricular node, then down the septum through the Bundle of His and through the endocardium. From there it propagates outward through the mid-myocardium to the epicardium, resulting in even ventricular contraction. Figure from Bers (2001)

the AP travels down the septum through the Purkinje fibers of the Bundle of His and along the endocardium. Purkinje fibers are specialized cells which allow rapid conduction. From here, the AP propagates from ventricular cell to ventricular cell outward though the mid-myocardium to the epicardium resulting in an even contraction of both ventricles.

The chapter from here will concentrate primarily upon the typical ventricular AP which initiates the process of contraction within these work horse cells of the heart.

4.1 The Resting Potential

The voltage gradient across the cardiac cell depends primarily upon the concentration gradient of ions across the membrane and upon the permeability of the membrane to each ion.

The dependence upon the concentration gradient is defined by the Nernst Equation:

$$E_m = \frac{RT}{zF} \ln \frac{[S]_o}{[S]_i} \tag{1}$$

where R is the gas constant, T in temperature in degrees Kelvin, z is the valence of the ion and F is Faraday's constant. The calculated membrane

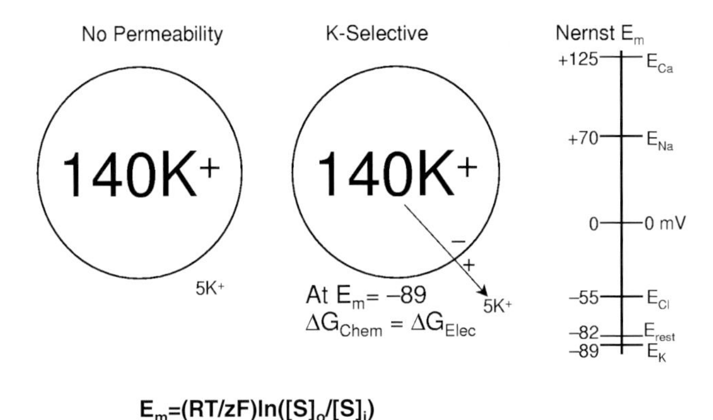

Fig. 3. The membrane voltage across a membrane will move to the reversal potential for an ion when the membrane becomes permeant to that ion. The reversal potential (or Nernst potential) is the voltage which exactly counteracts the chemical gradient for the ion across the membrane such that no net ion flux is produced. The reversal potentials for various relevant ions at physiological concentrations is shown on the scale on the right. Figure from Bers (2001)

voltage (E_m) is the reversal potential of the ion, S, at the concentrations $[S]_o$ and $[S]_i$ outside and inside the cell, respectively. The reversal potential is the voltage which must be applied to the cell membrane to exactly counteract the chemical gradient for S across the membrane such that no net ion transfer across it will take place through an open channel.

For instance, the reversal potential of K at the intracellular concentration of 140 mM and 5 mM extracellular concentration is −89 mV (Fig. 3). This means that if a voltage of −89 mV exists across the membrane, no net K will be transported across the membrane through an open K channel. If the voltage is above −89 mV, K will flow outward and if it is below −89 mV, K will flow inward.

Note the effect of ion flow upon the voltage in this system. If one starts at a potential above −89 mV, as K ions flow outward, positive ions are removed from the cellular interior, thus making it more negative and decreasing E_m. This process will continue until the system achieves the reversal potential where ion flow will cease.

Physiologically, however, we must deal with more than just one ion. Concentrations gradients of Cl, Na, and Ca also exist across the membrane and each of these ions influences the observed membrane voltage. These influences depend upon the reversal potential for each ion at its physiological concentration and the current which each ion produces as it traverses the membrane. The higher the permeability of the membrane to each ion, the higher the current produced by that ion. The higher the current, the more each ion leaving or entering the cell will decrease or increase the membrane voltage toward its reversal potential. Thus at any instance in time the observed E_m will nearly always be determined physiologically by numerous flowing currents generated by the transfer of a number of different ions at once.

Consider the cardiac myocyte at its resting potential of −82 mV. This membrane voltage is close to the reversal potential for K across the membrane (−89 mV). If K were the only ion flowing, the membrane voltage would be exactly −89 mV and no current would flow. But because other ions such as Na and Ca are flowing in small amounts across the membrane at this voltage, they tend to "pull" the voltage toward their respective reversal potentials. Nevertheless, the membrane is far more permeant to K than these other ions, thus the resting voltage, while not exactly the reversal potential for K, is very close to it and only a small amount of outward K current persistently flows at rest.

The membrane at rest is most permeant to K because the dominant open channel at the resting potential is the inward rectifier (I_{K1}). The behavior of this channel is characterized in Fig. 4B. The plots in this figure are known as current-voltage or IV curves. The data for IV curves are generated by performing an electrophysiological technique known as patch clamp.

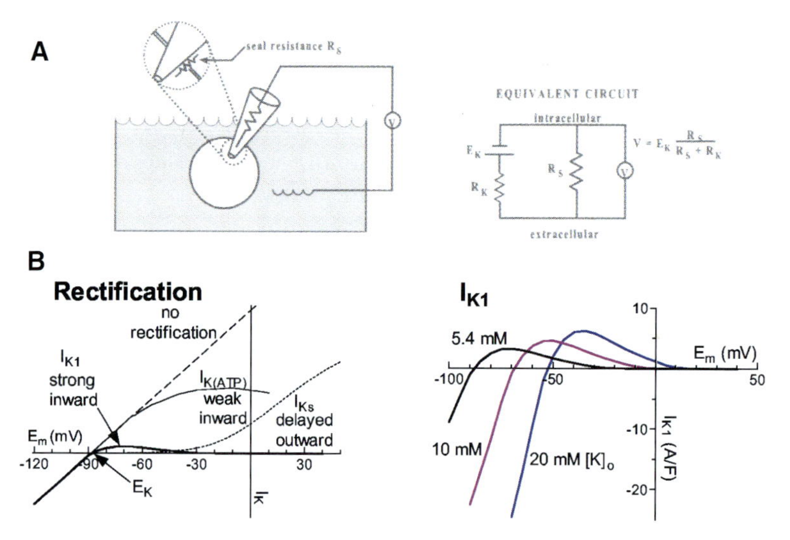

Fig. 4. (**A**) A diagram of a typical voltage clamp experiment (*left*). A glass pipette is sealed to the cell membrane, followed by the application of suction which breaks the membrane and allows electrical access to the cell. The diagram on the right represents the circuit which is formed upon rupture of the patch. (Diagrams from The Axon Guide for Electrophysiology and Biophysics Laboratory Techniques, Axon Instruments, Inc.) (**B**) The IV relationship for I_{K1}. The channel is a strong inward rectifier, preferring to conduct inward current as opposed to a channel with no rectification (*dashed line, left*). The panel on the right shows the shift in reversal potential which takes place at higher $[K]_o$. Figure from Bers (2001)

A diagram which demonstrates whole cell patch clamp is depicted in Fig. 4A. A glass pipette is sealed to the surface of the cell and the membrane is broken by applying suction to the tip. Thus electrical access to the cell is attained and an electrical circuit is established through the cell membrane (Fig. 4A, right). Current can then be injected into the circuit to maintain the membrane at a given voltage. By blocking or inactivating all other relevant currents, the current passing through an individual group of channels can be measured at any membrane voltage.

I_{K1} is an rectifying current. When current through a normal open K channel is passed, it falls upon the dotted line in Fig. 4B(left). It passes positive, outward current when E_m is above the reversal potential of $-89\,\mathrm{mV}$ and negative, inward current when E_m is below this value. However, I_{K1} is smaller than expected at voltages positive to the reversal potential. Thus the channel is inwardly rectifying i.e. it prefers to pass inward current.

4.2 Construction of the AP Waveform

As the AP is conducted to the cell in vivo, current is injected into the cell. This causes a depolarization to the threshold potential (Fig. 5A). The threshold for cardiac ventricular myocytes is that voltage which begins to open fast Na channels.

Fast Na channels, like most of the channels involved in the composition of the AP waveform, are activated by changes in E_m. This activation can be thought of as the opening of a charged gate which moves in response to the change in voltage. Similarly, all E_m-dependent channels exhibit deactivation which is simply the reverse of activation. Both of these processes are described in Fig. 5B. As can be seen, the fast Na channel has a voltage dependence of activation which is relatively negative. Hence, it takes relatively little depolarization from the resting potential to start to open them.

As Na channels open and the permeability of the SL for Na increases, the permeability of K through I_{K1} decreases (Fig. 4B) and the membrane voltage tends to rise toward the Na reversal potential. As the voltage increases, the permeability of the membrane for Na increases until the Na channels are maximally activated (Fig. 5B), resulting in a steep increase in voltage (Fig. 5A). Also note that as the voltage increases more and more, the other channels in Fig. 5B open. These channels have a relatively fast activation rate and with depolarization fast Na current (I_{Na}), T-type Ca current ($I_{Ca,T}$) and LTCC current (I_{Ca}) will progressively activate.

Many E_m-dependent channels also exhibit inactivation, which is a separate process from (de)activation. Inactivation is often voltage-dependent and the effects can be observed as a decrease in current following activation upon sustained membrane depolarization (e.g. Ca channels in Fig. 6B).

After the rapid depolarization phase of the AP, Na channels rapidly inactivate, decreasing Na permeability. At the same time, Ca channels, primarily Ca current (I_{Ca}), open and current through the Na-Ca exchanger, which brings

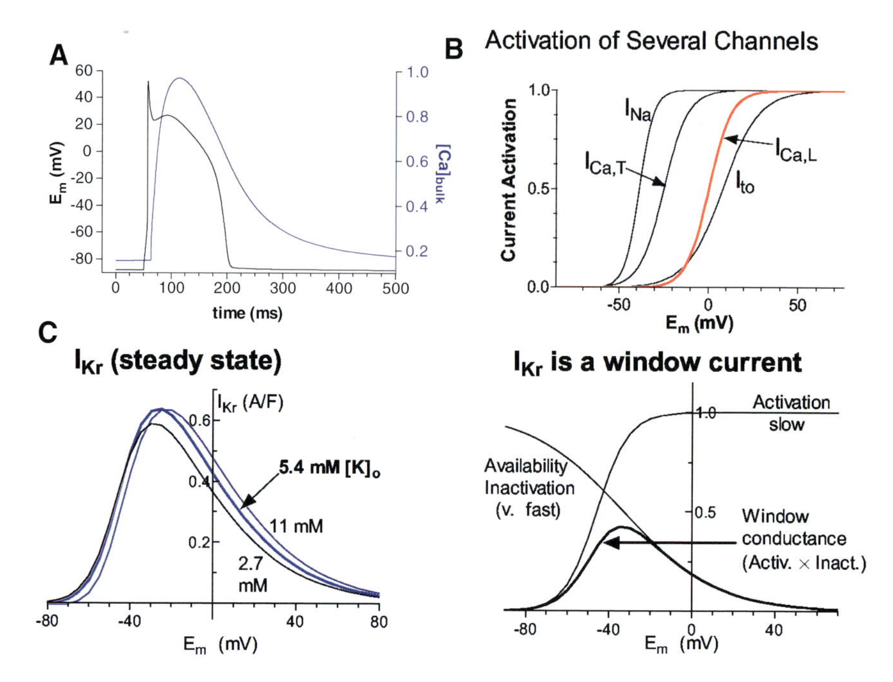

Fig. 5. (**A**) The AP with an associated $[\mathrm{Ca}]_i$ transient. (**B**) The activation of relevant fast currents as a function of voltage. The order of activation during the rapid AP depolarization is roughly from left to right as the voltage increases. (**C**) The IV relationship for I_{Kr}. This current normally operates in the "window" underneath the curves. Figures from Bers (2001)

in three Na for one Ca (see below) begins to appear. These currents tend to sustain depolarization. At the same time, K channels also begin to open, namely I_{TO} and I_{Kr}. Each of these outward currents activates in order and tends to repolarize the membrane toward the K reversal potential. Thus a balance is achieved which results in the plateau phase of the AP in Fig. 5A.

The steady-state IV curve for I_{Kr} is shown in Fig. 5C. This relationship results from the activation and inactivation vs. voltage relationships in Fig. 5D. Note that the channel has a tendency to activate as voltage goes up. However, because the rate of this activation is relatively slow, the current does not appear during the very fast depolarization phase of the AP (Fig. 5A). Instead, the current gradually appears during the plateau phase of the AP. As the current becomes stronger, the voltage declines further toward the reversal potential for K. Note that the voltage decline gradually relieves the inactivation of the channel during the sustained depolarization, making the current even stronger and further increasing the permeability to K. The region under the activation and inactivation relationships is known as a window current. The

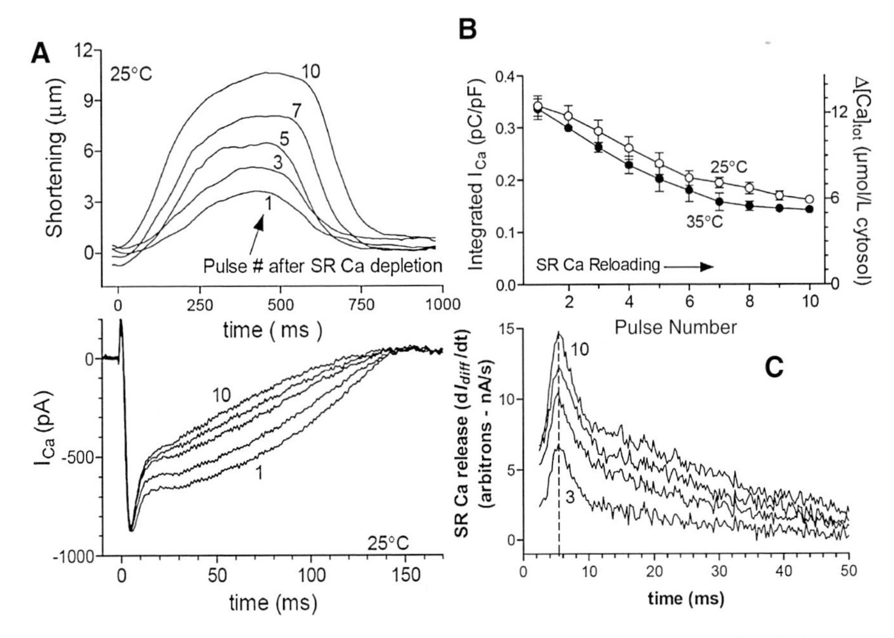

Fig. 6. Ca-dependent inactivation with increased SR Ca release (Puglisi et al., 1998). (**A**) As the SR is reloaded and SR Ca release increases with each post-caffeine stimulation, the degree of Ca-dependent inactivation increases. (**B**) The decrease in the amount of Ca transported in through I_{Ca} with each stimulation at room temperature and body temperature. (**C**) The rate of Ca-dependent inactivation with each stimulation. This rate is an indication of the release rate of Ca into the junction

current will persist as long as the voltage is held in this region and will never completely inactivate.

Figure 7 shows the currents involved in the AP over time. A second slow K current, I_{Ks}, appears during reploarization as the voltage continues to decline. Eventually, I_{K1} activates (Fig. 4) to bring E_m below the optimum for I_{Kr} and I_{Ks} and the membrane voltage returns to the resting potential and I_{K1} is once again, the only major K current flowing.

5 Ca Channels

There are two classes of Ca channels in cardiac muscle, T-type and L-type. The LTCC are characterized by large conductances, slower activation and higher voltage of activation relative to T-type channels (Fig. 5B, Fig. 8B). Though T-type current is more prominent in pacemaker cells and in atrial cells (Hagiwara et al., 1988; Bean, 1989) and may become significant in ventricular cells in disease states such as chronic heart failure (Nuss and Houser, 1993), it is either modest (Mitra and Morad, 1986) or not detectable (Bean, 1989; Nuss and Houser, 1993) in normal ventricular myocytes.

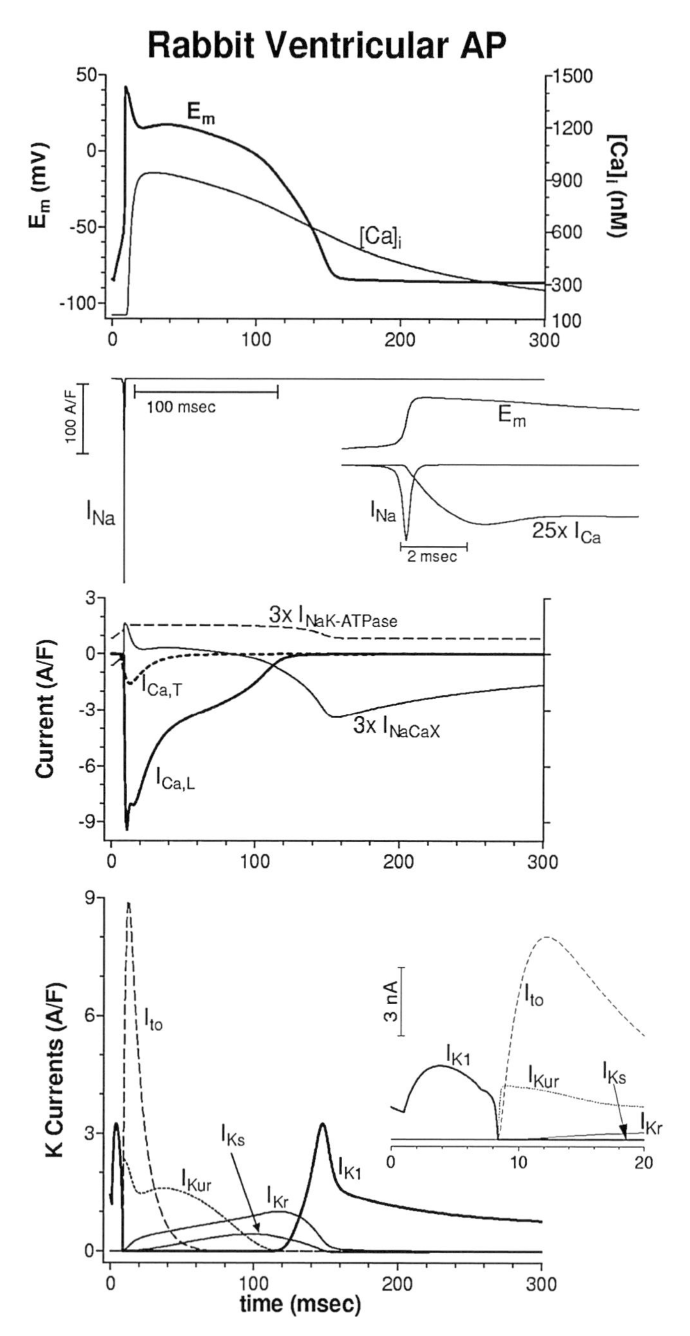

Fig. 7. The rabbit ventricular action potential and its associated currents. The voltage at any instance in time is a function of the reversal potential and the permeability of the membrane to each ion. Figure from Bers (2001)

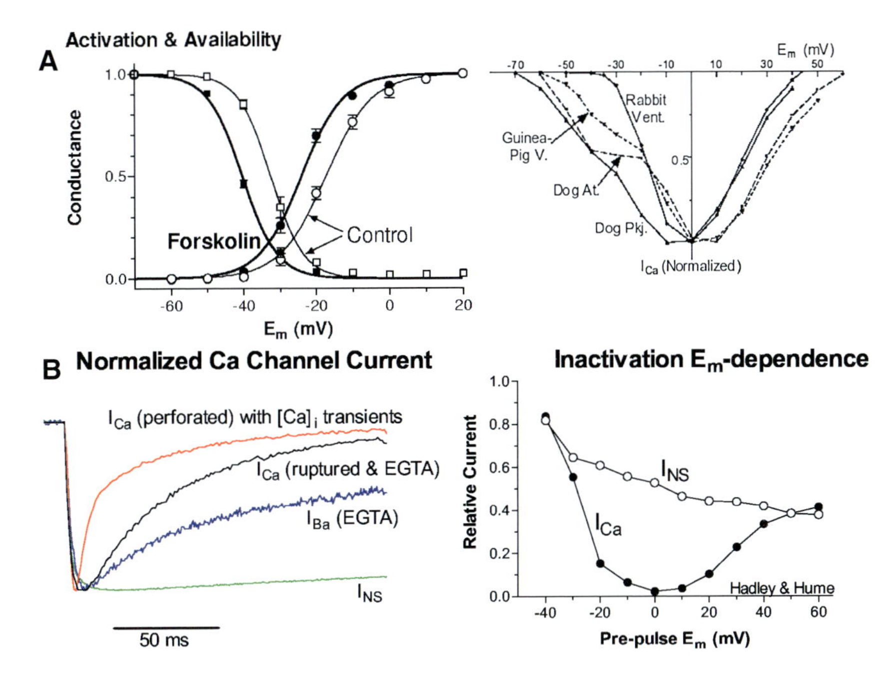

Fig. 8. (**A**) Steady-state activation and inactivation (availability) curves (*right*) and the IV relationship for I_{Ca} in various species (*left*). The activation and inactivation curves shift upon protein kinase A stimulation similar to what would result from binding of an adrenergic agonist like norepinephrine. (**B**) Ca-dependent inactivation of I_{Ca} under different conditions (*right*). The *left hand* figure shows the degree of inactivation as a function of voltage when Ca is passed through the channel as opposed to when no Ca in present (I_{NS}) (Hadley and Hume, 1987)

The LTCC, on the other hand, is ubiquitous in cardiac myocytes. It has voltage-dependent inactivation (Fig. 8A) which, when combined with its steady-state activation, results in a very small window current which is usually not significant. Activation of protein kinase A, usually through adrenergic receptor binding, results in a shift in both availability curves as well as a large increase in conductance.

Voltage dependent inactivation of this current is relatively slow. However, it has a Ca-dependent inactivation in the intact cell which results in transient kinetics (Hadley and Hume, 1987, Fig. 8B, left). Figure 8B demonstrates this Ca-dependence. It shows the magnitude of the current at the end of a 500 ms depolarization. When the channel is only allowed to pass Na and K, which are also conducted by this channel in the absence of Ca, much of the current still remains. However, when the channel passes Ca, as it normally does, the inactivation is nearly complete because the Ca binds to the channel and causes the current decline.

As stated above, Ca influx from the LTCC into the junctional cleft is the principal stimulus for CICR in the cardiac myocyte. This local release of Ca is graphically demonstrated in Fig. 6 by tracking the degree of Ca-dependent inactivation as a function of the amount of Ca released into the cleft. The Ca currents in Fig. 6A (bottom) are those which are evoked during contractions after emptying of the SR with caffeine (see below). The first post-caffeine contraction and SR Ca release are small (Fig. 6A, top) and as a result, the Ca-dependent inactivation following the peak of the I_{Ca} is relatively minor (Fig. 6C). However, as more Ca enters the cell and is taken up into the SR with each subsequent depolarization, the Ca current inactivates more rapidly and to a greater extent due to Ca-dependent inactivation.

6 The SR Ca Release Channel

The SR Ca release channel is more commonly referred to as the ryanodine receptor (RyR) because of its high affinity for the plant alkaloid, ryanodine. Indeed, it was through its association with this ligand that it was isolated (Inui et al., 1987). The protein exists in 3 isoforms. RyR 3 was isolated from brain and appears to be expressed only to a small extent in mammalian skeletal muscle. RyR 1 is the primary mammalian skeletal muscle isoform and RyR 2 is the isoform of ventricular muscle.

The RyR is a homotetramer and is extremely large with a size of 2.3 megadaltons (Saito et al., 1988; Wagenknecht et al., 1989, Fig. 9A). The complex is 28 nm along each side and 14 nm in height. Indeed, the complex spans much of the 15 nm gap between the SR and the SL bring it into close proximity to the L-type Ca channel. This closeness makes Ca mediated coupling between the LTCC and the RyR which results in CICR more efficient.

Figure 9B shows some key regulatory domains of the RyR. Indeed, the cardiac RyR is, in reality, a megacomplex including FKBP12.6, a protein kinase A anchoring protein (mAKAP) and two phosphatases (PP1 and PP2A) (Marx et al., 2000). The protein also interacts with calmodulin, junctin and triadin.

The protein exhibits both Ca-dependent activation and inactivation from the cytosolic side (Fig. 10A). The Ca-dependence of the opening can be demonstrated in numerous ways. For instance, Ca release is graded with I_{Ca} magnitude (Beuckelmann and Wier, 1988, Fig. 10B) and this is considered to be a defining characteristic of cardiac ECC. The Ca-dependence of the channel can also be demonstrated when it is incorporated in planar lipid bilayers into which the RyR have been incorporated (Xu et al., 1998). The current through single channels can be measured in response to changes in Ca on the *cis* (i.e. the cytosolic) side of the membrane (Fig. 11A & B). Indeed, SR Ca release can even be measured as the rate of ^{45}Ca efflux from isolated SR membrane vesicles (Meissner and Henderson, 1987, Fig. 11B).

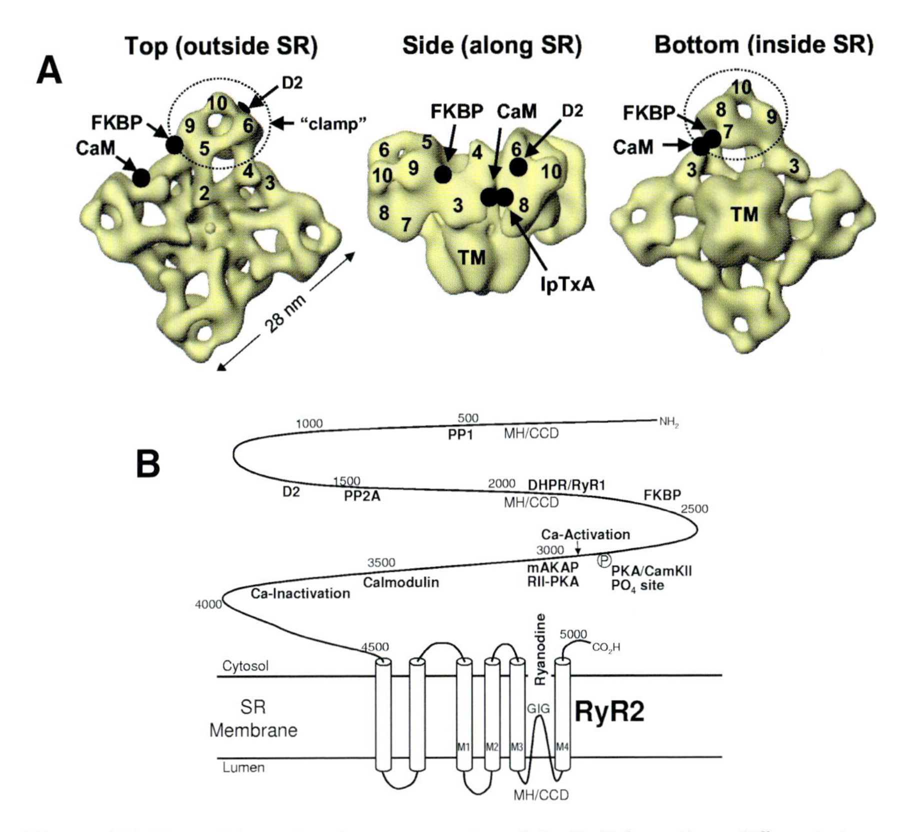

Fig. 9. (**A**) Three-Dimensional reconstruction of the RyR from three different views. Various domains are indicated: TM, transmembrane region; IpTxA, imperitoxin A; CaM, calmodulin; D2, divergency region 2; FKBP, FK506 binding protein. (**B**) Domains in the cardiac RyR sequence. The 4 transmembrane domains M1-M4 are indicated and there may be 2 more. Approximate locations along the primary structure of several sites of either interaction (e.g. phosphatases 1 and 2A, PP1 and PP2A; mAKAP, kinase anchoring protein), a putative pore region (GIG), PKA/CaMKII phosphorylation site (P) and Ca effector sites are also indicated. A few sites important in RyR1 are also shown, e.g. mutation sites associated with malignant hyperthermia or central core disease (MH/CCD) and sites where skeletal muscle DHPRs may interact. Figures both appear in Bers (2001)

Figure 12 shows a popular technique also to evaluate SR Ca release, measurement of Ca sparks. A "Ca spark" is a release event from a single release unit in a cell, probably one junction. Sparks are typically measured through the use of a confocal microscope which images the cell through a plane about 1 μm thick while excluding light from outside this plane. Usually the myocyte is kept under a condition where release events are expected to be relatively sparse and easy to see in isolation without interference from simultaneous fir-

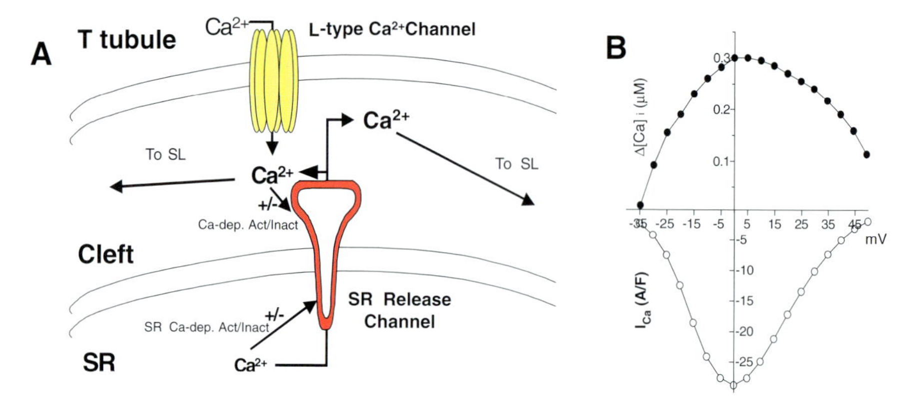

Fig. 10. (**A**) Diagram illustrating the functional effects of Ca upon the RyR. (**B**) Model data representing a typical relationship between I_{Ca} and SR Ca release (i.e. $\Delta[\mathrm{Ca}]_i$) and E_m. Figures are from Shannon et al. (2004)

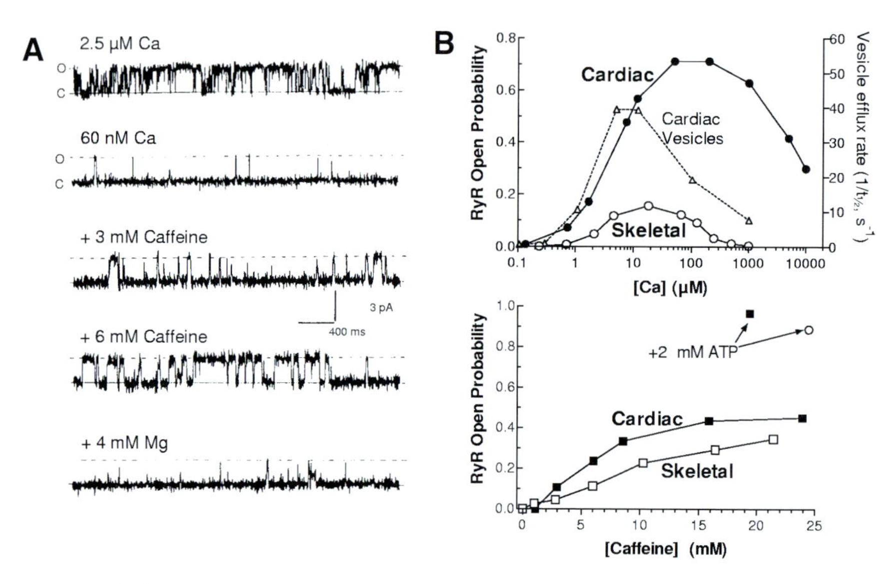

Fig. 11. (**A**) Caffeine, Mg and [Ca]-dependence of RyR gating (channels incorporated into lipid bilayers). Single cardiac Ca release channel records from Rousseau and Meissner (1989) show that lowering cis- (cytosolic) [Ca] reduces channel opening (B, o=open c=closed), that caffeine activates the channel at low [Ca] and that Mg blocks the channel. (**B**) Ca dependence of channel open probability (P_o) as in A for cardiac and skeletal Ca release channel (Xu et al., 1998) or of the rate of ^{45}Ca efflux from cardiac SR vesicles (Meissner and Henderson, 1987, upper). The *lower panel* shows the effect of caffeine on cardiac and skeletal RyR in bilayers. Addition of ATP caused the channels activated by caffeine to become almost fully open (Smith et al., 1986; Rousseau et al., 1988)

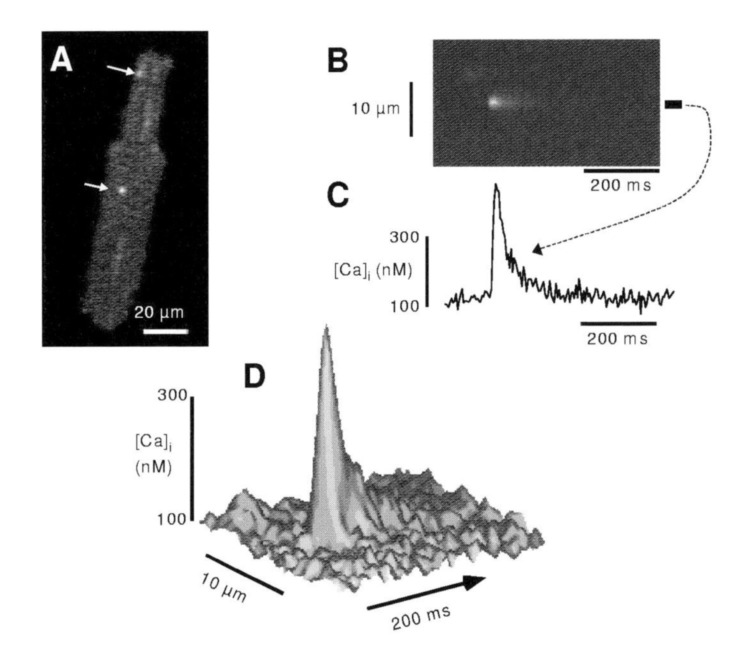

Fig. 12. Ca sparks in an isolated mouse ventricular myocyte. (**A**) Two dimensional laser scanning confocal fluorescence image of a myocyte loaded with the Ca-sensitive indicator fluo-3, exhibiting two Ca sparks (*arrows*). (**B**) Line scan image along the long axis of the myocyte (only part is shown). Scans were repeated every 4 ms and stacked from *left* to *right*. Distance along the cell is in the vertical direction. (**C**) Line graph of $[Ca]_i$ at the spot indicated by the bar in B (1 µm). (**D**) Surface plot of $[Ca]_i$ during a Ca spark, indicating the temporal and spatial spread of Ca. Figure as in Bers (2001)

ing of neighboring release units. By evaluating the size (i.e. the extent of the increase in Ca-dependent fluorescence) and frequency of these events, SR Ca release may be evaluated under these conditions.

7 Ca-Induced Ca Release

Figure 13A shows seven hypotheses for mechanisms by which CICR might take place. It is beyond the scope of this chapter to address each of these mechanisms. See Bers (2001) for an excellent review. Suffice it to say that Ca influx through I_{Ca} into the junctional space is nearly universally believed to be the major mechanism by which Ca is brought into the cell to stimulate release from the SR with the others playing a minor or modulatory role.

On the surface, CICR might be expected to be highly regenerative. Ca release from one RyR would be expected to cause release from its neighbors which would, in turn, cause release from their neighbors until release

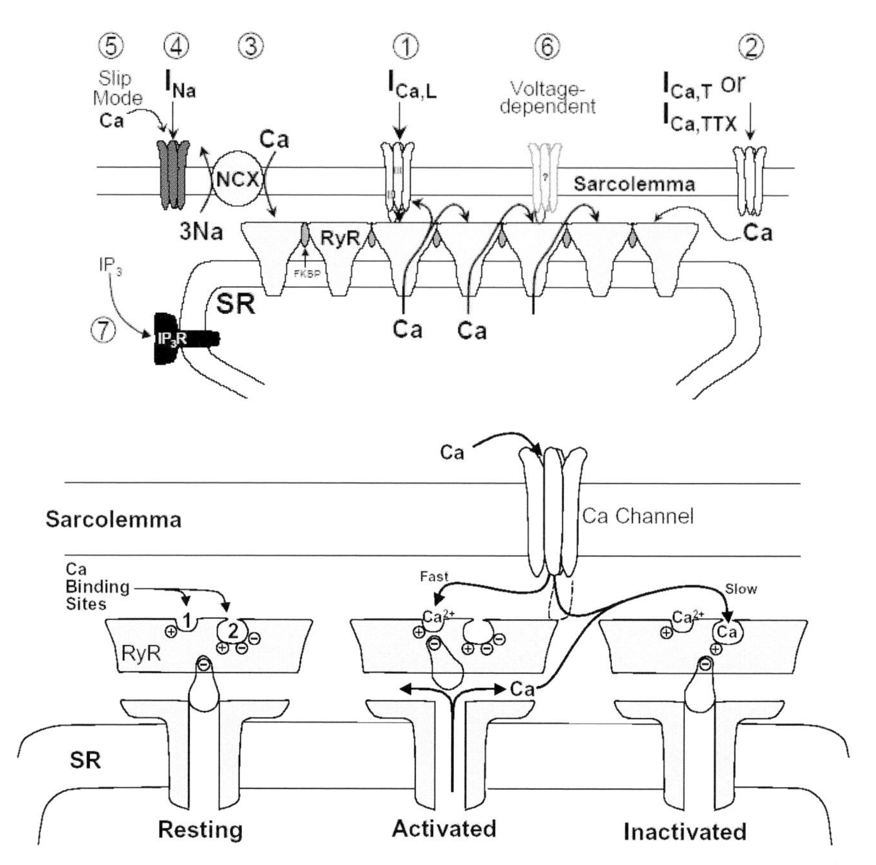

Fig. 13. (**A**) Potential ECC mechanisms in cardiac muscle. **1.** L-type Ca Current. **2**: T-type or TTX sensitive Ca current (ICa,TTX). **3.** Ca influx via Na/Ca exchange driven directly by Em-dependence of $I_{Na/Ca}$. **4.** Ca influx via Na/Ca exchange driven by local high $[Na_i]$, secondary to I_{Na}. **5.** Altered selectivity of Na channels (allowing Ca permeation) with PKA activation. **6.** Voltage-dependent SR Ca release (VDCR) **7.** Inositol (1,4,5)-trisphosphate (IP3)-induced SR Ca release (IP3ICR). (**B**) Diagram of CICR (Fabiato, 1983). From the resting state (channel closed), Ca may bind rapidly to a relatively low affinity site (1), thereby activating the RyR. Ca may then bind more slowly to a second higher affinity site (2) moving the release channel to an inactivated state. As cytoplasmic [Ca] decreases, Ca would be expected to dissociate from the lower affinity activating site first and then more slowly from the inactivating site to return the channel to the resting state. Figure from Bers (2001)

becomes an all-or-none event. However, each of the techniques demonstrated in Fig. 11B results a bell shaped relationship between release and stimulatory [Ca]. The rising phase demonstrates the action of Ca-dependent activation of the RyR while the falling phase is a result of Ca-dependent inactivation. The existence of the Ca-dependent inactivation may help to explain why CICR is not all-or-none. Figure 13B shows one classical hypothesis which may explain why this is so (Fabiato, 1983). Ca influx causes a rise in Ca in the junctional space. this Ca binds to low affinity, but rapidly binding RyR activation sites and the channel opens. The Ca then binds to higher affinity but slower binding sites which inactivate the channel. Hence Ca both opens and closes the channel.

The effects of Ca upon the RyR on the lumenal side of the SR membrane may also contribute to the turn off of SR Ca release. SR Ca release should increase with increasing SR Ca simply because the gradient for Ca across the membrane increases. However, SR Ca may also affect the process of release itself. Figure 14A shows data which suggests the existence of such effects in the intact cell. SR Ca release was measured as a function of SR [Ca] (Shannon et al., 2000). An increase in SR Ca gradient would be expected to increase the gain (Ca released per I_{Ca} Ca stimulus) linearly and the fractional release (released Ca/SR [Ca]) would be expected to be constant. In contrast to these expectations, gain and fractional release both increased nonlinearly. A potential explanation for such an effect is demonstrated in Fig. 14B. The plot is of the results of a bilayer experiment in which the *cis* [Ca] ($[\mathrm{Ca}]_i$) dependent opening was tested with high and low [Ca] on the trans (lumenal) side on the membrane (Györke and Györke, 1998). At higher lumenal [Ca], the probability of opening was higher over the range of $[\mathrm{Ca}]_i$. Free SR Ca ($[\mathrm{Ca}]_{\mathrm{SR}}$), therefore, tends to promote activation of the channel.

Figure 14C presents a recent hypothesis which would at least partially explain such an effect (Gyorke et al., 2004). According to this model, triadin and/or junctin tend to promote opening of the RyR when bound tightly to it. Ca free calsequestrin would tend to bind to triadin and/or junctin, thus loosening the bond with the RyR which results in a tendency to keep the RyR closed.

8 Na-Ca Exchange

As Ca is released, the concentration spikes high in the junctional cleft and diffuses out underneath the SL membrane and finally into the cytosol. Relaxation takes place as the Ca is transported out of the cytosol either back into the SR or into the extracellular space. The primary mode of transport of Ca out of the cell is through the Na-Ca exchanger.

The Na-Ca exchanger operates to transport one Ca ion in one direction in exchange for three Na in the opposite direction. Ordinarily during cellular relaxation, it transports Ca out and Na in, using the Na transport down its

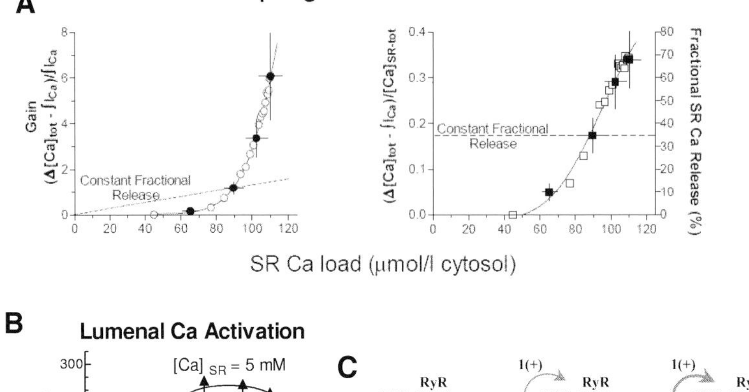

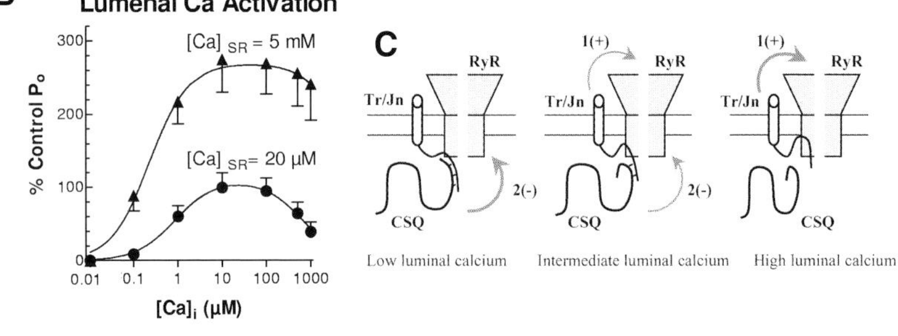

Fig. 14. (**A**) Gain of SR Ca release and fractional release as a function of SR Ca (Shannon et al., 2000). The relationships are steeply nonlinear at higher SR Ca indicating that SR Ca is affecting the RyR. The *dashed lines* indicate the expected result when SR Ca does not affect the process of release. (**B**) Probability of opening of the RyR in artificial bilayers at high and low *trans* [Ca] (i.e. $[Ca_{SR}]$). (**C**) The hypothesized functional interactions between calsequestrin, junctin, triadin, and RyR in the cardiac SR membrane as proposed by Gyorke et al. (2004). At low lumenal Ca concentrations, calsequestrin (CSQ) inhibits the RyR channel complex by competing with the RyR for triadin and junctin (A). At high lumenal Ca this inhibition is relieved as Ca binding leads to less CSQ-triadin/junctin interaction and RyR channel activity is stimulated (Gyorke et al., 2004)

gradient across the SL membrane to drive Ca out of the cell from low cytosolic concentration to high extracellular concentration. This is known as operation in the "forward mode". However the process of Na-Ca exchange is reversible. The role of reverse mode Na-Ca exchange in bring Ca into the cell to modulate CICR is controversial and won't be discussed here.

Na-Ca exchange activity is sometimes measured in isolated SL vesicles (Fig. 15). Vesicles with high Na concentration in the lumen may be diluted into an extravesicular environment with low Na concentration. This results in uptake of radioactive ^{45}Ca tracer into the vesicles. This process may be reversed by diluting the vesicle back onto a relatively high Na environment.

Because the exchange is three Na for one Ca, there is a net charge moved across the membrane in the direction of Na transport. The exchanger is

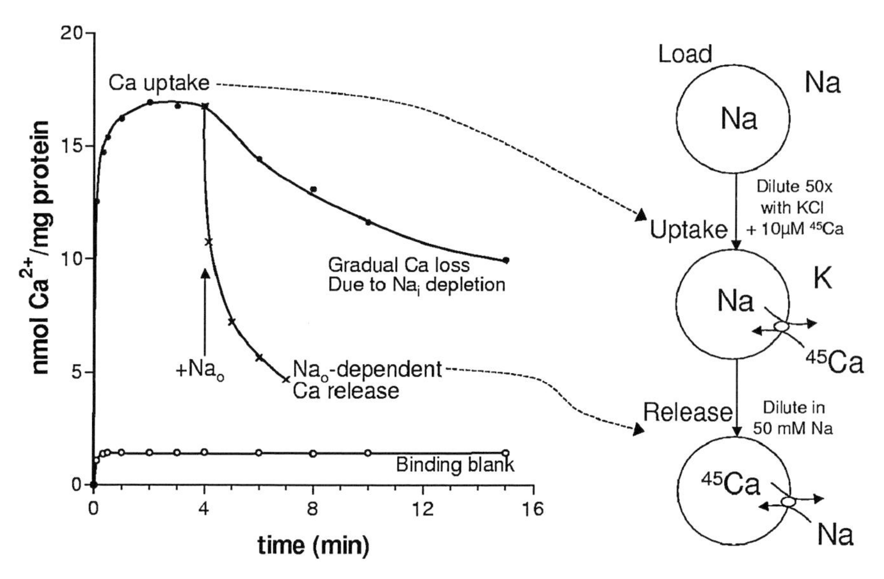

Fig. 15. A typical Na-Ca exchange experiment using Na-loaded SL vesicles. ^{45}Ca tracer is transported into the vesicles upon dilution into a low Na extravesicular environment. Dilution back into a high Na environment reverses the process. Figure from Bers (2001)

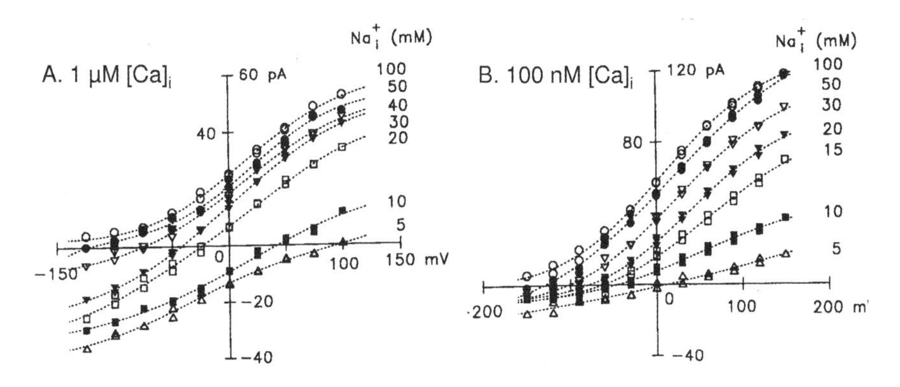

Fig. 16. Na/Ca current-voltage relationships with constant $[\mathrm{Ca}]_o$ (2 mM), $[\mathrm{Na}]_o$ (150 mM) (Matsuoka and Hilgemann, 1992). $[\mathrm{Na}]_i$ was varied from 5 to 100 mM. The $[\mathrm{Ca}]_i$ of 1 µM in the *right panel* is close to the peak systolic value whereas the $[\mathrm{Ca}]_i$ of 100 nm in the *left panel* is close to the normal diastolic value

therefore electrogenic and E_m affects the transport. Figure 16 shows 2 families of IV curves for Na-Ca exchange. Because the transport is dependent upon both Na and Ca, the reversal potential of the exchanger shifts with changes in concentration of either or both ions:

$$E_{Na/Ca} = 3E_{Na} - 2E_{Ca} \tag{2}$$

where E_{Na} and E_{Ca} are the reversal potentials for Na and Ca, respectively. The deviation of $E_{Na/Ca}$ from E_m during the AP is a measure of the driving force for transport and the current generated by the exchanger (Fig. 17). When $E_m > E_{Na/Ca}$, Ca influx (reverse mode Na-Ca exchange) is favored. When $E_m < E_{Na/Ca}$ during most of the AP, Ca efflux through forward mode exchange is favored. Note that a relatively small change in $[Na]_i$, as might take place in a ventricular myocyte in chronic heart failure, can result in a relatively large change in $E_{Na/Ca}$. This can dramatically affect transport in both directions.

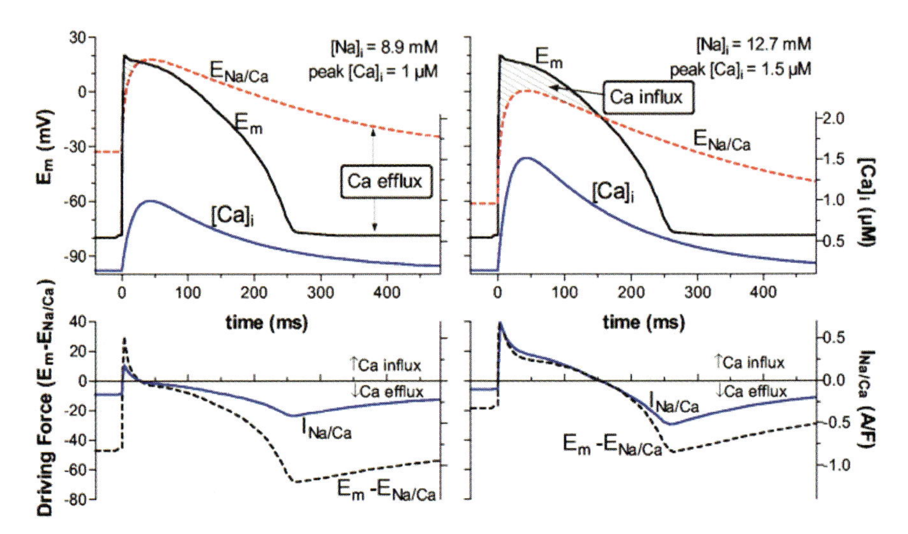

Fig. 17. Theoretical changes in $E_{Na/Ca}$ and $I_{Na/Ca}$ during an AP. When $E_m > E_{Na/Ca}$, Ca influx via the Na/Ca exchanger is thermodynamically favored (*shaded areas*) and the further the deviation away from E_m, the greater the driving force and the current in that direction. When $E_m < E_{Na/Ca}$, Ca extrusion is favored. Resting $[Ca]_i = 150\,nM$, $[Ca]_o = 2\,mM$ and $[Na]_o = 140\,mM$ for all traces. Figure from Bers (2001)

The Na-Ca exchanger is also regulated allosterically via binding of Ca and Na to the cytosolic side of the protein (Hilgemann et al., 1992a,b). The physiological functions of $[Ca]_i$-dependent activation and $[Na]_i$-dependent inactivation of the exchanger are not clear. $[Ca]_i$-dependent activation may stimulate the exchanger when Ca is high during systole and turn it off as Ca declines during diastole to prevent excessive loss of Ca. The $[Na]_i$-dependent inactivation may act to prevent Ca overload via reverse mode exchange when Na within the cytosol is high.

9 SL Ca Pump

The SL Ca pump is the second known mechanism for Ca extrusion from the cell. The pump uses ATP as an energy source to pump Ca up its concentration gradient. The SL Ca pump has a V_{max} of only about 2 µmol/l cytosol/s and is not thought to contribute a great deal to the $[Ca]_i$ decline during diastole (Bassani et al., 1995). However, data from Dixon and Haynes (1989) (Table 3) indicate that the pump may be stimulated by both protein kinase A and calmodulin-dependent protein kinase. Therefore, the SL Ca pump activity should be considered under specific conditions where its activity may be expected to be high.

Table 3. Data taken from Dixon and Haynes (1989)

Kinetic Properties of the Cardiac Sarcolemmal Ca-Pump

	V_{max} nmol/mg pn/min	V^*_{max} µmol/L cytosol/sec	K_m(Ca) nM	n(Hill)
Basal	1.7 ± 0.3	0.28	1800 ± 100	1.6 ± 0.1
+PKA	3.1 ± 0.5	0.50	1000 ± 100	1.7 ± 0.1
+Calmodulin	15.0 ± 2.5	2.43	64 ± 1.4	3.7 ± 0.2
+PKA-calmodulin	36.0 ± 6.5	5.83	63 ± 1.7	3.7 ± 0.1

10 SR Ca Pump

The transporter which is largely responsible for moving Ca from the cytosol back into the SR is the sarco(endo)plasmic reticulum Ca ATPase (SERCA) or the SR Ca pump. Two striated muscle SERCA proteins have been sequenced and cloned. One of these has two isoforms and is from fast twitch skeletal muscle. They are SERCA1a, predominant in adults, and SERCA1b found in neonates. The other protein also has several isoforms. SERCA2a is found in slow twitch skeletal muscle and cardiac muscle and SERCA2b and a third isoform, SERCA3, are found largely in non-excitable cells.

The SR Ca pump transports 2 Ca ions into the organelle for each ATP hydrolyzed (Reddy et al., 1996). Lower stoichiometries are often reported in vitro, probably because of pump-independent leak of Ca back out of the SR vesicles or because of contaminating ATPases.

Unlike skeletal muscle, the cardiac SR contains the regulatory protein phospholamban (PLB). PLB is a small integral membrane protein (22 kDa) and is an endogenous inhibitor of the SR Ca pump. It decreases Ca transport by decreasing its affinity for Ca without altering the V_{max}. PLB inhibits both SERCA1a and SERCA2a but not SERCA3. It is present at concentrations

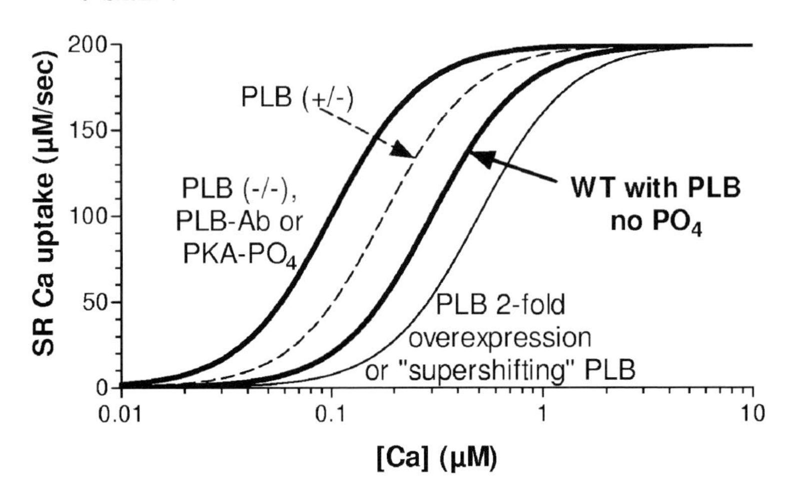

Fig. 18. The influence of PLB on Ca transport. Over-expression of PLB shows a shift in the K_m for Ca transport to a higher value. Knockout (-/-) of the PLB results in a shift in the opposite direction and more SR Ca uptake. PLB antibody (PLB-Ab) and phosphorylation of PLB (PLB-PO_4) both evoke the same response by preventing PLB binding to the SR Ca pump. Figure from Bers (2001)

comparable to the SR Ca pump, itself, in cardiac myocytes (Tada et al., 1983).

Figure 18 shows the influence of PLB on Ca transport. Over-expression of PLB shows a shift in the K_m for Ca transport to the right resulting in lower uptake at the same [Ca] over the physiological range. Knockout (i.e. elimination) of the PLB results in a leftward shift. This effect is mimicked by PLB antibody and phosphorylation of PLB, both of which prevent PLB binding to the SR Ca pump.

The effect of phosphorylation is particularly important. It is one means by which β-adrenergic agonists accelerate relaxation in the heart (Li et al., 2000) and it likely increases the efficiency of the pump as well (Frank et al., 2000; Shannon et al., 2001). The accelerated uptake biases the competition for Ca in favor of the SR Ca pump resulting in increased SR [Ca] and release (see below).

Like most enzymes, the SR Ca pump can work in both directions. Consider a usual reversible enzyme and the way that it operates. Initially, the enzyme begins to convert substrate to product. However, as the product concentration increases, the reversal of this process begins to take place and product begins to be converted back into substrate. The forward reaction continues, however, the net conversion of substrate to product is slowed due to the back conversion of some of the product back to substrate. Eventually the product concentration builds and the reaction comes to equilibrium. This is the state at which substrate conversion to product equals the product conversion to substrate such that the net conversion rate is zero.

The SR Ca pump is no different than the generic enzyme described above:

$$[\mathrm{Ca}]_i + \mathrm{ATP} \rightleftharpoons [\mathrm{Ca}]_{\mathrm{SR}} + \mathrm{ADP} + \mathrm{PO}_4$$

$[\mathrm{Ca}]_i$ and ATP are the substrates, ADP, PO_4 and $[\mathrm{Ca}]_{\mathrm{SR}}$ are the products. As product builds during diastole, a unidirectional flux back out of the SR through the pump begins.

It is important to note that, unlike our generic enzyme, this reaction never comes to equilibrium. In the intact cell, the forward direction is always greater than the reaction in the reverse direction. The net pump rate is always forward bringing Ca into the SR. This net forward rate is equal to the leak rate of Ca back out of the SR. Consider Fig. 19. If the pump operates in the forward direction without any reversal flux, it operates in the forward mode at approximately 20% of the $V_{\max}$ (right). Under this condition, the leak is equal to the uptake or about 21 μmol/l cytosol/s. If however, a reversal flux of 20.7 μmol/l cytosol exists, the leak is only 0.3 μmol/l cytosol/s. The majority of this SR Ca leak is likely due to diastolic release of Ca through the RyR. The exact amount of it under physiological conditions is unknown but like CICR, this SR Ca leak is strongly SR [Ca] dependent (Shannon et al., 2002) and may be increased in heart failure (Shannon et al., 2003).

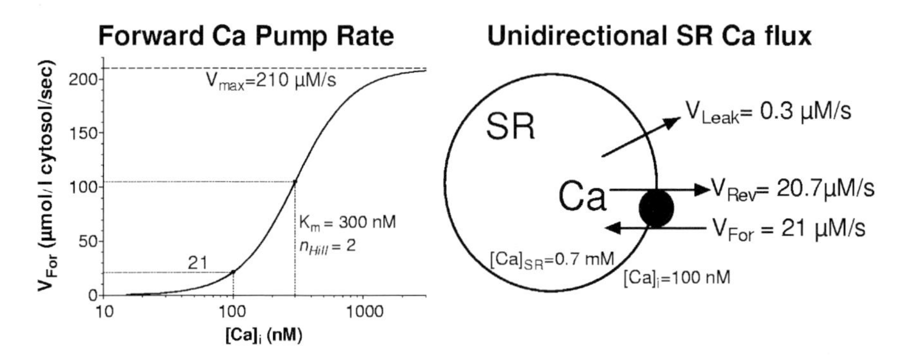

Fig. 19. The *left panel* demonstrates the uptake rate of Ca with no reversal of the SR Ca pump using reasonable parameters and a simple Hill equation ($V_{\mathrm{for}} = V_{\max}/(1 + K_m/[\mathrm{Ca}]_i{}^n)$). At 100 nM $[\mathrm{Ca}]_i$, the rate is 21 μmol/l cytosol/s. On the *right* the figure demonstrates the uptake when reversal flux is accounted for. In the presence of a relatively low leak flux (V_{Leak}, 0.3 μM/l cytosol/s) the forward rate remains high with the remainder of the efflux from the SR accounted for by reversal flux (V_{rev}, 20.7 μmol/l cytosol/s). Figure from Bers (2001)

A likely scenario for SR Ca uptake during the cardiac cycle is demonstrated in Fig. 20. As release takes place, $[\mathrm{Ca}]_{\mathrm{SR}}$ (the product) decreases while $[\mathrm{Ca}]_i$ (the substrate) increases. The result is an increase in the forward rate and a decrease in the reverse rate of the pump. Thus a sudden increase in net uptake is observed in the figure. As uptake proceeds, $[\mathrm{Ca}]_{\mathrm{SR}}$ builds resulting

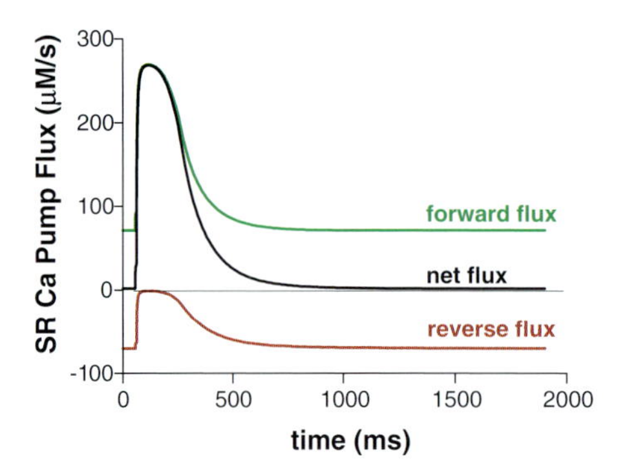

Fig. 20. The SR Ca uptake during the cardiac cycle. During release, $[Ca]_{SR}$ decreases and $[Ca]_i$ increases resulting in decreased reverse flux and increased forward flux. As $[Ca]_i$ declines and $[Ca]_{SR}$ increases, forward flux falls and reversal flux increases until the difference (net flux) becomes equal to the leak flux

in an increase in the reversal flux. At the same time $[Ca]_i$ decreases resulting in a decrease in the forward flux. This process continues until the net flux is equal to the leak flux.

11 The Balance of Fluxes

The SR Ca pump, the Na-Ca exchanger, the SL Ca pump and mitochondrial uptake are the transporters responsible for Ca removal from the cytosol. Mitochondrial uptake is considered to be relatively minor and is not considered in detail in this chapter. Each of these transporters competes for Ca in the cytosol. Critical to any description of Ca dynamics within the ventricular myocyte is the proper balance of competition for the Ca substrate by these transporters during diastolic Ca decline.

Figure 21 shows the percentage of Ca transported by each method of transport. In a normal rabbit myocyte (similar to the human) most of the Ca is transported by the SR Ca pump back into the SR (Bassani et al., 1994; Puglisi et al., 1996, 70%). Of the remaining 30%, 28% is transported by the Na-Ca exchanger with the remainder leaving through the SL Ca pump or taken up into the mitochondria. Note, however, that this balance varies with species. The rat and mouse, for instance, take up much more of the Ca which leaves the cytosol back up into the SR with much less leaving through the Na-Ca exchanger.

Why worry about the balance of fluxes? There are several reasons for this. The first is that this balance is an indication of the state of ECC. At steady-state, the Ca entering the cytosol is equal to the amount of Ca which leaves with each beat. By the same token, the amount of Ca uptake into the SR is equal to the amount of release. Therefore the Ca uptake divided by the

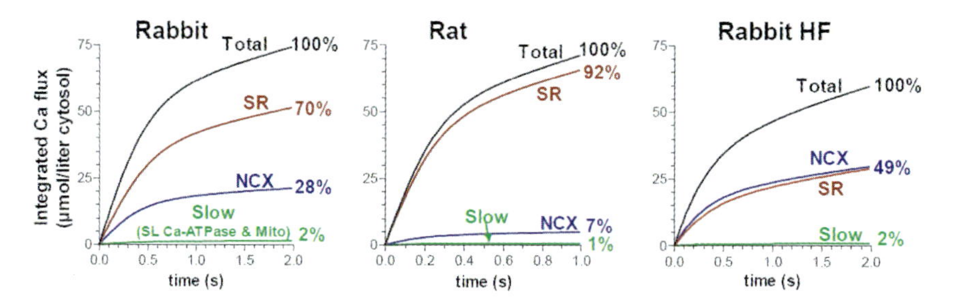

Fig. 21. The competition between the methods of Ca transport out of the cytosol. The "slow" transport is the sum of the mitochondrial uptake and the SL Ca pump transport. The balance of these fluxes changes with species which express the transporters at different relative levels (Bassani et al., 1994; Puglisi et al., 1996). It is also altered in pathological conditions such as heart failure (Pogwizd et al., 1999) where the SR Ca pump is down regulated and the Na-Ca exchanger is up regulated

amount which leaves via the SL Ca pump and the Na-Ca exchanger is a rough indicator of the gain of ECC (i.e. the amount of Ca release/the amount of Ca stimulus through I_{Ca}).

In addition, the balance of fluxes is also altered in heart failure where much more of the Ca leaves via the Na-Ca exchanger (Pogwizd et al., 1999). Indeed, if one wishes to accurately describe any physiological or pathophysiological effect which alters Ca transport, it is critical that this balance be correct. Any alteration of transport will affect not only the Ca decline, but indirectly the balance of Ca influx and release, as well.

12 Myocyte Contraction

I will conclude with a brief consideration of the way in which the $[\text{Ca}]_i$ transient, the generation of which we have just described, results in the generation of force. The myocyte contracts in response to increases in cytosolic [Ca]. This response is graded over the physiological range (Fig. 22). Most experimental results have been obtained in muscle preparations which have had their SL removed either mechanically or chemically in what are commonly referred to as "skinned" preparations. Under these conditions, the myofilaments are directly exposed to the [Ca] in the bath and the steady-state force which is generated in response to this [Ca] can be measured. This response varies with species and condition.

The protein within the myofilament complex which binds Ca is Troponin C (TnC). At the core of this complex is a long thin filament made up of a continuous string of monomers known as the actin filament. TnC is a major buffer for Ca within the cytosol, binding more Ca than any other Ca binding protein during the $[\text{Ca}]_i$ transient (Table 2). In the relaxed state, TnC lies in loose

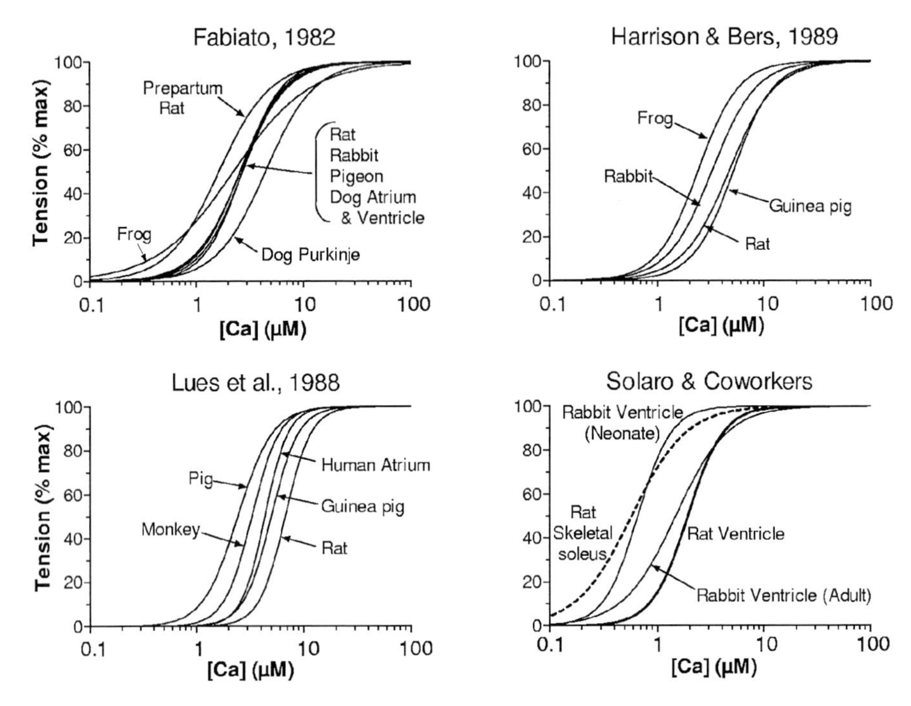

Fig. 22. Myofilament Ca sensitivity in cardiac muscle preparations. The data is replotted from the references indicated (Fabiato, 1982; Lues et al., 1988; Harrison and Bers, 1989; Wattanapermpool et al., 1995; Solaro and Rarick, 1998). Figure from Bers (2001)

association with Troponin I (TnI, Fig. 23). TnI is an inhibitory subunit which serves as a regulatory protein within the complex. TnI, through an intermediary protein, Troponin T (TnT), is associated with tropomycin. Tropomycin lies along the actin filament, covering about seven actin monomers. Together all of the troponin molecules form the troponin complex and these also are located at every seventh actin monomer.

Force within the myofilament complex is generated by association of a thick filament made principally of myosin molecules with the thin filament. Each thick filament has about 300 myosins, each of which has a head which binds to and forms cross bridges with the actin monomers. These heads also contain sites for the ATP hydrolysis which drives the reaction.

The troponin complex and tropomycin stericly prevent myosin heads from interacting with actin (Fig. 23). In the resting state as stated above, TnI interacts near its amino end to the carboxy ends of both TnT and TnC. In the resting state at low Ca, the amino terminus of TnI also binds specifically to actin, thus preventing the myosin-actin crossbridge formation. When Ca binds to the amino end of TnC, it binds strongly to TnI, causing TnI dissociation from actin. This change in the association of TnC with TnI is sensed by TnT

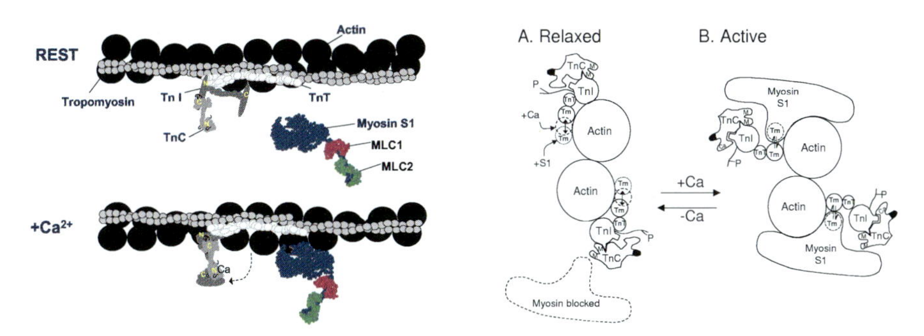

Fig. 23. The diagram depict the changes on a molecular scale which result in activation of the myofilaments. Tropomysin (Tm) lies along the chain of actin molecules covering the myosin binding sites. Upon Troponin C (TnC) binding to Ca, it association to troponin I (TnI) tightens. The complex pulls Tm axially through the Troponin T (TnT) intermediary. The movement exposes the myosin binding sites on the actin molecule which allows myosin-mediated contraction. Figure from Bers (2001)

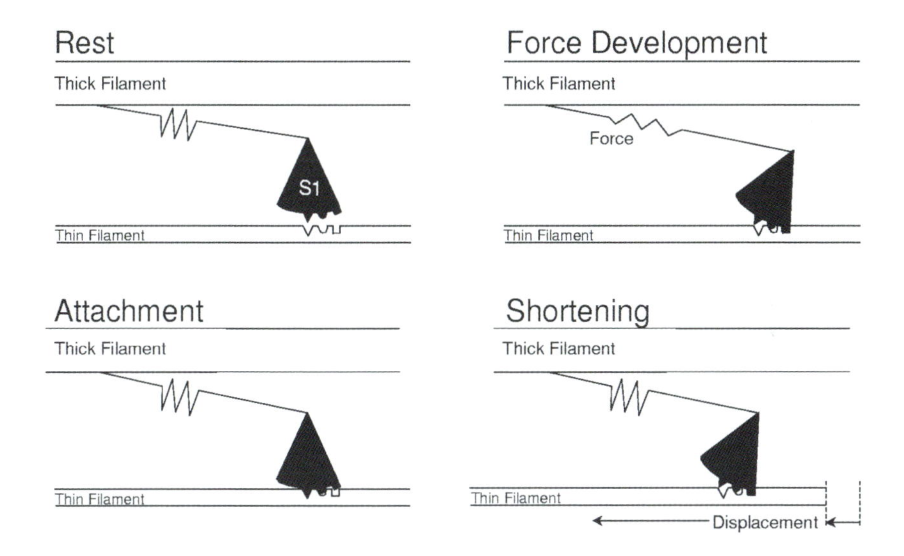

Fig. 24. A model of active crossbridges based on the original diagram by Huxley and Simmons (1971). The resting, detached crossbridge is represented by M·ATP in the chemical model in which actin is bound to ATP. The attached crossbridge prior to developing force has a hydrolyzed ATP associated with the head (A·M·ADP·P_i). As the attached crossbridge develops force, the phosphate (P_i) is released with the ADP still bound in a high energy state (A·M·ADP*). The head then rotates, and the filaments slide relative to one another (A·M·ADP and A·M). Figure from Bers (2001)

and causes movement of tropomycin axially, allowing myosin to interact with actin. When TnI is phosphorylated the rate of Ca dissociation from TnC increases resulting in more rapid relaxation of the myofilaments.

The basic sliding filament model for myofilament contraction was first proposed by Huxley and Simmons (1971, Fig. 24). Once the myosin head can bind to actin, a series of chemical steps results in the process of contraction (Goldman and Brenner, 1987) and these have been correlated with the physico-mechanical scheme. At rest, myosin (M) is either complexed with ATP (M·ATP) or with hydrolyzed ATP (M·ADP·P_i). As $[Ca]_i$ rises M·ADP·P_i is allowed to interact with actin at the myosin binding site. At that point, phosphate (P_i) is rapidly released and the acto-myosin passes into a high energy state (A·M·ADP*). The transition through this state and A·M·ADP may make up the so called "power stroke". This is is the stage of contraction in which the myosin head rotates, thus generating force and moving the actin filament in parallel along the myosin axis. Eventually ADP also is released (A·M) at which point ATP rapidly binds and induces dissociation of myosin from actin. This cycle continues until $[Ca]_i$ declines.

References

Bassani, J. W. M., Bassani, R. A., and Bers, D. M. (1994). Relaxation in rabbit and rat cardiac cells: Species-dependent differences in cellular mechanisms. *J. Physiol.*, 476:279–293.

Bassani, R. A., Bassani, J. W. M., and Bers, D. M. (1995). Relaxation in ferret ventricular myocytes: Role of the sarcolemmal Ca ATPase. *Eur J. Physiol.*, 430:573–578.

Bean, B. P. (1989). Classes of calcium channels in vertebrate cells. *Annu. Rev. Physiol.*, 51:367–384.

Bers, D. M. (2001). *Excitation-Contraction Coupling and Cardiac Contractile Force.* Kluwer Academic Publishers, Dordrecht, 2 edition.

Beuckelmann, S. J. and Wier, W. G. (1988). Mechanism of release of calcium from sarcoplasmic reticulum of guinea pig cardiac cells. *J. Physiol.*, 405:233–255.

Dixon, D. A. and Haynes, D. H. (1989). Kinetic charaterization of the Ca^{2+}-pumping ATPase of cardiac sarcolemma in four states of activation. *J. Biol. Chem.*, 264:13612–13622.

Fabiato, A. (1982). Calcium release in skinned cardiac cells: Variations with species, tissues, and development. *Fed. Proc.*, 41:2238–2244.

Fabiato, A. (1983). Calcium-induced release of calcium from the cardiac sarcoplasmic reticulum. *Am. J. Physiol.*, 245:C1–C14.

Frank, K., Tilgmann, C., Shannon, T., Bers, D., and Kranias, E. (2000). Regulatory role of phospholamban in the efficiency of cardiac sarcoplasmic reticulum Ca^{2+} transport. *Biochemistry*, 39(46):14176–82.

Goldman, Y. E. and Brenner, B. (1987). Special topic: Molecular mechanism of muscle contraction. general introduction. *Annu. Rev. Physiol.*, 49:629–636.

Györke, I. and Györke, S. (1998). Regulation of the cardiac ryanodine receptor channel by luminal Ca^{2+} involves luminal Ca^{2+} sensing sites. *Biophys. J.*, 75(6):2801–2810.

Gyorke, I., Hester, N., Jones, L. R., and Gyorke, S. (2004). The role of calsequestrin, triadin, and junctin in conferring cardiac ryanodine receptor responsiveness to luminal calcium. *Biophys. J.*, 86:2121–2128.

Hadley, R. W. and Hume, J. R. (1987). An intrinsic potintial-dependent inactivation mechanism associated with calcium channels in guinea pig myocytes. *J. Physiol.*, 389:205–222.

Hagiwara, N., Irisawa, H., and Kameyama, M. (1988). Contribution of two types of ca currents to the pacemaker potentials of rabbit sino-atrial node cells. *J. Physiol.*, 359:233–253.

Harrison, S. M. and Bers, D. M. (1989). Influence of temperature on the calcium sensitivity of the myofilaments of skinned ventricular muscle from the rabbit. *J. Gen. Physiol.*, 93:411–428.

Hilgemann, D. W., Collins, A., and Matsuoka, S. (1992a). Steady-state and dynamic properties of cardiac sodium-calcium exchange. secondary modulation by cytoplasmic calcium and atp. *J. Gen. Physiol.*, 100:933–961.

Hilgemann, D. W., Matsuoka, S., Nagel, G. A., and Collins, A. (1992b). Steady-state and dynamic properties of cardiac sodium-calcium exchange. sodium-dependent inactivation. *J. Gen. Physiol.*, 100:905–932.

Huxley, A. F. and Simmons, R. M. (1971). Proposed mechanism of force generation in striated muscle. *Nature*, 233:533–538.

Inui, M., Saito, A., and Fleischer, S. (1987). Isolation of the ryanodine receptor from cardiac sarcoplasmic reticulum and identity wih the feet structures. *J. Biol. Chem.*, 262:15637–15642.

Li, L., DeSantiago, J., Chu, G., Kranias, E. G., and Bers, D. M. (2000). Phosphorylation of phospholamban and troponin i in β-adrenergic-induced acceleration of cardiac relaxation. *Am. J. Physiol.*, 278:H769–H779.

Lues, I., Siegel, R., and Harting, J. (1988). Effect of isomazole on the responsiveness to calcium of the contractile elements in skinned cardiac muscle fibres of various species. *Eur. J. Pharmacol.*, 146:145–153.

Marx, S. O., Reiken, S., Hisamatsu, Y., Jayaraman, T., Burkhoff, D., Rosemblit, N., and Marks, A. R. (2000). PKA phosphorylation dissociates FKBP12.6 from the calcium release channel (ryanodine receptor): Defective regulation in failing hearts. *Cell*, 101:365–376.

Matsuoka, S. and Hilgemann, D. W. (1992). Steady-state and dynamic properties of cardiac sodium-calcium exchange. ion and voltage dependence of the transport cycle. *J. Gen. Physiol.*, 100:963–1001.

Meissner, G. and Henderson, J. S. (1987). Rapid calcium release from cardiac sarcoplasmic vesicles is dependent on Ca^{2+} and is modulated by Mg^{2+}, adenine nucleotide and calmodulin. *J. Biol. Chem.*, 262:3065–3073.

Mitra, R. and Morad, M. (1986). Two types of ca channels in guinea-pig ventricular myocytes. *Proc. Natl. Acad. Sci.*, 83:5340–5344.

Nuss, H. B. and Houser, S. R. (1993). T-type ca current is expressed in hypertrophied adult feline left ventricular myocytes. *Circ. Res.*, 73:777–782.

Page, E. (1978). Quantitative ultrastructural analysis in cardiac membrane physiology. *Am. J. Physiol.*, 235:C147–C158.

Page, E., Mccallister, L. P., and Power, B. (1971). Stereological measurements of cardiac ultrastructures implicated in excitation-contraction coupling. *Proc. Natl. Acad. Sci.*, 68:1465–1466.

Pogwizd, S. M., Qi, M., Yuan, W. Samarel, A. M., and Bers, D. M. (1999). Upregulation of Na^{+}/Ca^{2+} exchanger expression and function in an arrhythmogenic rabbit model of heart failure. *Circ. Res.*, 85(11):1009–1019.

Puglisi, J. L., Bassani, Rosana, A., Bassani, J. W. M., Amin, J. N., and Bers, D. M. (1996). Temperature and relative contributions of Ca transport systems in cardiac myocyte relaxation. *Am. J. Physiol.*, 270:H1772–H1778.

Puglisi, J. L., Yuan, W., Bassani, J. W. M., and Bers, D. M. (1998). Ca influx through Ca channels in rabbit ventricular myocytes during action potential clamp: Influence of temperature. *Circ. Res.*, submitted.

Reddy, L. G., Jones, L. R., Pace, R. C., and Stokes, D. L. (1996). Purified, reconstituted cardiac Ca^{2+} -ATPase is regulated by phospholamban but not by direct phosphorylation with Ca^{2+}/calmodulin-dependent protein kinase*. *J. Biol. Chem.*.

Rousseau, E., Ladine, J., Liu, Q. Y., and Meissner, G. (1988). Activation of the Ca^{2+} release channel of skeletal muscle sarcoplasmic reticulum by caffeine and related compounds. *Arch Biochem Biophys*, 265:75–86.

Rousseau, E. and Meissner, G. (1989). Single cardiac sarcoplasmic reticulum Ca^{2+}-release channel: Activation by caffeine. *Am. J. Physiol.*, 256:H328–H333.

Saito, A., Inui, M., Radermacher, M., Frank, J., and Fleischer, S. (1988). Ultrastructure of the calcium release channel of sarcoplasmic reticulum. *J. Cell Biol.*, 107:211–219.

Scriven, D. R., Dan, P., and Moore, E. D. (2000). Distribution of proteins implicated in excitation-contract coupling in rat ventricular myocytes. *Biophys. J.*, 79(5):2682–2691.

Shannon, T., Chu, G., Kranias, E., and Bers, D. (2001). Phospholamban decreases the energetic efficiency of the sarcoplasmic reticulum Ca pump. *J. Biol. Chem.*, 276(10):7195–201.

Shannon, T. R., Ginsburg, K. S., and Bers, D. M. (2000). Potentiation of fractional SR Ca release by total and free intra-SR Ca concentration. *Biophys. J.*, 78:334–343.

Shannon, T. R., Ginsburg, K. S., and Bers, D. M. (2002). Quantitative Assessment of the SR Ca^{2+} Leak-Load Relationship. *Circ. Res.* 91:594–600.

Shannon, T. R., Pogwizd, S. M., and Bers, D. M. (2003). Elevated sarcoplasmic reticulum Ca leak in intact ventricular myocytes from rabbits in heart failure. *Circ. Res.* 93:592–594.

Shannon, T. R., Wang, F., Puglisi, J., Weber, C. R., and Bers, D. M. (2004). A mathematical treatment of integrated Ca dynamics within the ventricular myocyte. *Biophys. J.*, 87:3351–3371.

Smith, J. S., Coronado, R., and Meissner, G. (1986). Single channel measurements of the calcium release channel from skeletal muscle sarcoplasmic reticulum. *J. Gen. Physiol.*, 88:573–588.

Soeller, C. and Cannell, M. B. (1997). Numerical simulation of local calcium movements during l-type calcium channel gating in the cardiac diad. *Biophys. J.*, 73(1):97–111.

Solaro, R. J. and Rarick, H. M. (1998). Troponin and tropomyosin: Proteins that switch on and tune in the activity of cardiac myofilaments. *Circ. Res.*, 83:471–480.

Tada, M., Inui, M., Yamada, M., Kadoma, M. A., Kuzuya, T., Abe, H., and Kakiuchi, S. (1983). Effect of phospholamban phosphorylation catalyzed by adenosine 3';5'-monophosphate and calmodulin-dependent protein kinases on calcium transport ATPase of cardiac sarcoplasmic reticulum. *J. Mol. Cell. Cardiol.*, 15:335–346.

Trafford, A. W., Diaz, M. E., O'Neill, S. C., and Eisner, D. A. (1995). Comparison of subsarcolemmal and bulk calcium concentration during spontaneous calcium release in rat ventricular myocytes. *J. Physiol.*, 488(577–586).

Wagenknecht, T., Grassucci, R., Frank, J., Saito, A., Inui, M., and Fleischer, S. (1989). Three-dimensional architecture of the calcium channel/foot structure of sarcoplasmic reticulum. *Nature*, 338:167–170.

Wattanapermpool, J., Reiser, P. J., and Solaro, R. J. (1995). Troponin I isoforms and differential effects of acidic ph on soleus and cardiac myofilaments. *Am. J. Physiol.*, 268:C323–C330.

Weber, C. R., Piacentino, V., Ginsburg, K. S., Houser, S. R., and Bers, D. M. (2002). Na^{+}-Ca^{2+} exchange current and submembrane [Ca^{2+}] during the cardiac action potential. *Circ. Res.*, 90:182–189.

Xu, L., Tripathy, A., Pasek, D. A., and Meissner, G. (1998). Potential for pharmacology of ryanodine receptor calcium release channels. *Ann. NY Acad. Sci*, 853:130–148.

Mechanisms and Models of Cardiac Excitation-Contraction Coupling

R.L. Winslow[1], R. Hinch[2] and J.L. Greenstein[1]

[1] Center for Cardiovascular Bioinformatics and Modeling and The Whitaker Biomedical Engineering Institute, The Johns Hopkins University School of Medicine and Whiting School of Engineering
[2] Mathematical Institute, University of Oxford

1 Introduction

Intracellular calcium (Ca^{2+}) concentration plays an important regulatory role in a number of cellular processes. Cellular influx of Ca^{2+} activates intracellular signaling pathways that in turn regulate gene expression. Studies have identified over 300 genes and 30 transcription factors which are regulated by intracellular Ca^{2+} [1,2]. Fluctuation of intracellular Ca^{2+} levels is also known to regulate intracellular metabolism by activation of mitochondrial matrix dehydrogenases. The subsequent effects on the tri-carboxylic acid cycle increase the supply of reducing equivalents (NADH, $FADH_2$), stimulating increased flux of electrons through the respiratory chain [3]. Most importantly, Ca^{2+} is a key signaling molecule in excitation-contraction (EC) coupling, the process by which electrical activation of the cell is coupled to mechanical contraction and force generation.

The purpose of this chapter is to review the important role of Ca^{2+} in cardiac EC coupling, with a particular focus on presentation of quantitative models of the EC coupling process. One of the most remarkable features of cardiac EC coupling is that structural and molecular properties at the microscopic scale have a profound influence on myocyte function at the macroscopic scale. Thus, modeling of cardiac EC coupling is confronted with the enormous challenge of needing to integrate diverse information collected at multiple scales of biological analysis into comprehensive models. In this chapter, we will consider three different approaches to such multi-scale modeling of cardiac EC coupling, reviewing the strengths and weaknesses of each approach.

2 The Molecular and Structural Basis of Cardiac EC Coupling

2.1 Structural Basis of EC Coupling

The nature of EC coupling is linked closely to both the micro-anatomical structure of the myocyte as well as the arrangement of contractile proteins within the cell. These relationships are shown in Fig. 1. The basic unit of contraction in the cardiac myocyte is the sarcomere. Individual sarcomeres are approximately 2.0 μm in length and are bounded on both ends by the Z-disks. The H-Zone contains the thick (myosin) filaments and is the region within which there is no overlap with thin (actin) filaments. The A-band is the region spanned by the length of the thick filaments. The shaded region in Fig. 1 represents the region of overlap of thick and thin filaments. Muscle contraction is accomplished by the sliding motion of the thick and thin filaments relative to one another in this region in response to elevated levels of intracellular calcium (Ca^{2+}) and ATP hydrolysis.

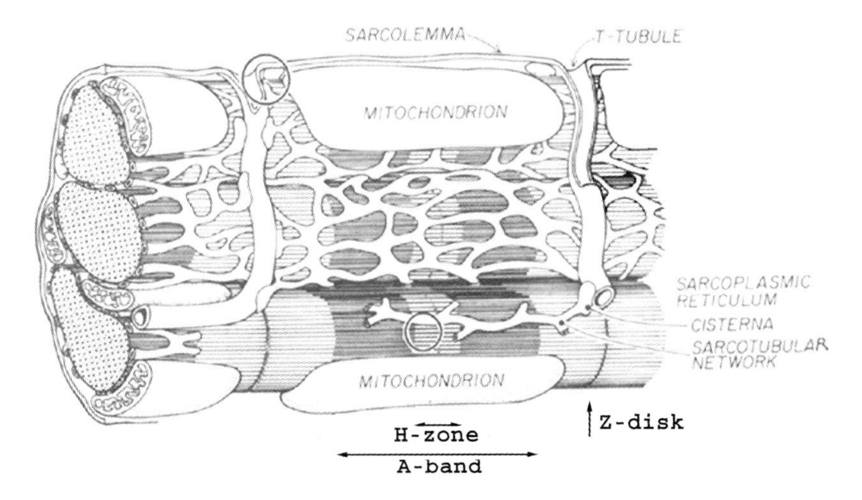

Fig. 1. Ultrastructure of the cardiac ventricular myocyte, illustrating the organization of sarcomeres and the T-tubules. Reproduced from Fig. 1.14 of Katz [73] with permission

Figure 1 also shows that sarcomeres are bounded on each end by the T-tubules [4]. T-tubules are cylindrical invaginations of the sarcolemma that extend into the myocyte (Fig. 2), approaching an organelle known as the sarcoplasmic reticulum (SR). The SR is comprised of two components known as junctional SR (JSR) and network SR (NSR). The NSR is a lumenal organelle extending throughout the myocyte. NSR membrane contains a high concentration of the SR Ca^{2+}-ATPase (SERCA2α) pump, which transports Ca^{2+}

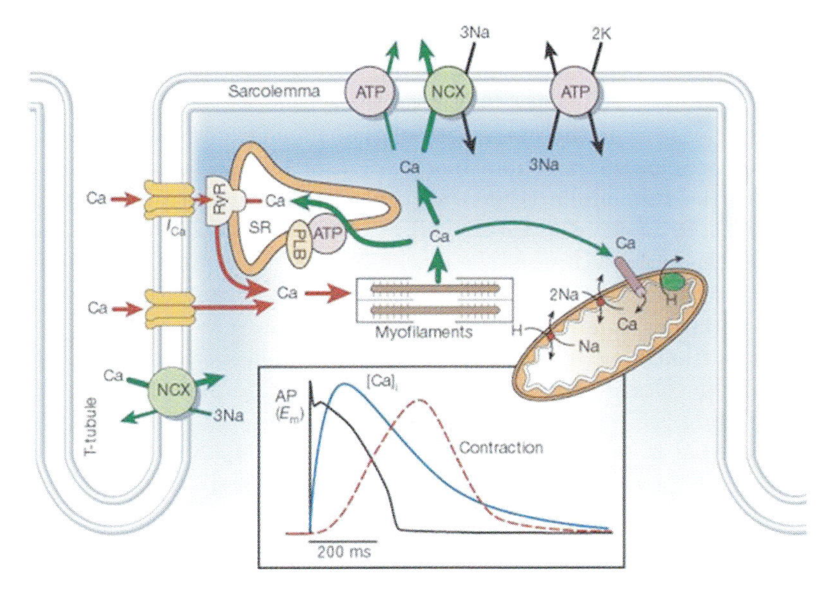

Fig. 2. Illustration of the physical organization and channel localization in the cardiac dyad. Reproduced from Bers Fig. 1 [66] with permission

from the cytosol into the lumen of the NSR (Fig. 2). The JSR is that portion of the SR most closely approximating the T-tubules. The close proximity of these two structures (~10 nm) forms a restricted region commonly referred to as the "dyad" or "subspace" with an approximate diameter of 400 nm. Ca^{2+} release channels (known as ryanodine receptors or RyRs) are located in the dyadic region of the JSR membrane. In addition, sarcolemmal L-Type Ca^{2+} channels (LCCs) are located preferentially within the dyadic region of the T-tubules, where they are in close opposition to the RyRs. It has been estimated that there are approximately 10 LCCs per dyad [5] and ~50,000 active LCCs per myocyte.

2.2 The Molecular Basis of Cardiac EC Coupling

2.2.1 Properties of LCCs

Cardiac LCCs are multi-subunit protein complexes located in the T-tubular membrane (for review, see [6]). The main functional subunit of the cardiac LCC is the α_{1c} subunit. Each α_{1c} subunit has 6 transmembrane segments (S1-S6) and four subunits assemble to create the LCC pore. LCCs undergo voltage-dependent activation and inactivation. The voltage sensor is located in the S4 segment and the inactivation gate is located in the S6 segment [7]. LCCs are also composed of auxiliary subunits. Four distinct β subunits, with multiple splice variants, bind to the channel complex to modify its properties [8]. An $\alpha_2\delta$ subunit also interacts with and modifies channel properties. Both the α_{1c}

and the β_{2a} subunit (the dominant β subunit isoform in heart) contain protein kinase A (PKA) phosphorylation sites [9].

LCCs also undergo inactivation in response to elevated Ca^{2+} levels within the dyadic space near the inner pore of the channel – a process referred to as Ca^{2+}-dependent inactivation. Inactivation of LCCs is mediated by Ca^{2+} binding to the calmodulin (CaM) molecule which is constitutively tethered to the LCC [10]. Following Ca^{2+} binding, CaM binds to an IQ-like domain located in the carboxyl tail of the α_{1c} subunit. This in turn leads to a conformational change of the EF hand region located downstream from the IQ domain, initiating channel inactivation [11].

An important issue, to be discussed in subsequent sections, is the relative magnitude of voltage- versus Ca^{2+}-dependent inactivation of LCCs. Recent data regarding the relative contributions of these processes are presented in Fig. 3 (adapted from Linz and Meyer, Fig. 11 [12]). Open triangles show the fraction of LCCs not voltage-inactivated and filled circles show the fraction of LCCs not Ca^{2+}-inactivated. Measurements were made in isolated guinea pig ventricular myocytes in response to an action potential (AP) clamp stimulus (upper panel). These data show that the fraction of LCCs not Ca^{2+}-inactivated is small relative to the fraction not voltage-inactivated. Thus, Ca^{2+}-inactivation dominates over voltage-inactivation during the time course of an AP.

Figure 3B shows recent data from the laboratory of David Yue and colleagues (adapted from Fig. 5 of Peterson et al. [10]). The left panel of Fig. 3B shows current through LCCs in response to a voltage-clamp step from a holding potential of $-80\,\mathrm{mV}$ to $-10\,\mathrm{mV}$. Two current traces are shown, one with Ca^{2+} and one with Ba^{2+} as the charge carrier. Ca^{2+}-dependent inactivation of LCCs is thought to be ablated when Ba^{2+} is used as the charge carrier. The Ba^{2+} current thus reflects properties of voltage-dependent inactivation of LCCs. Comparison of the two current traces suggests that over a time scale of a few hundred milliseconds, Ca^{2+}-dependent inactivation is fast and strong whereas voltage-dependent inactivation is slow and weak. The right panel of Fig. 3B shows results obtained when a mutant CaM incorporating point mutations which prevent Ca^{2+} binding is over expressed in the myocytes, thus definitively ablating Ca^{2+}-dependent inactivation. Under this condition, the Ca^{2+} and Ba^{2+} currents are similar, thus providing an independent measurement confirming that Ca^{2+}-dependent inactivation of LCCs is fast and strong, whereas voltage-dependent inactivation is slow and weak.

2.2.2 Properties of RyRs

Ryanodine receptors (RyRs) are ligand-gated channels located in the dyadic region of the JSR membrane. These channels open following binding of Ca^{2+} in response to elevated Ca^{2+} levels in the dyad, thus releasing Ca^{2+} from the JSR. Cardiac RyRs are composed of four 565-kDa subunits and four 12-kD FK506 binding proteins (FKBP12.6). FKBP12.6 stabilizes the closed state

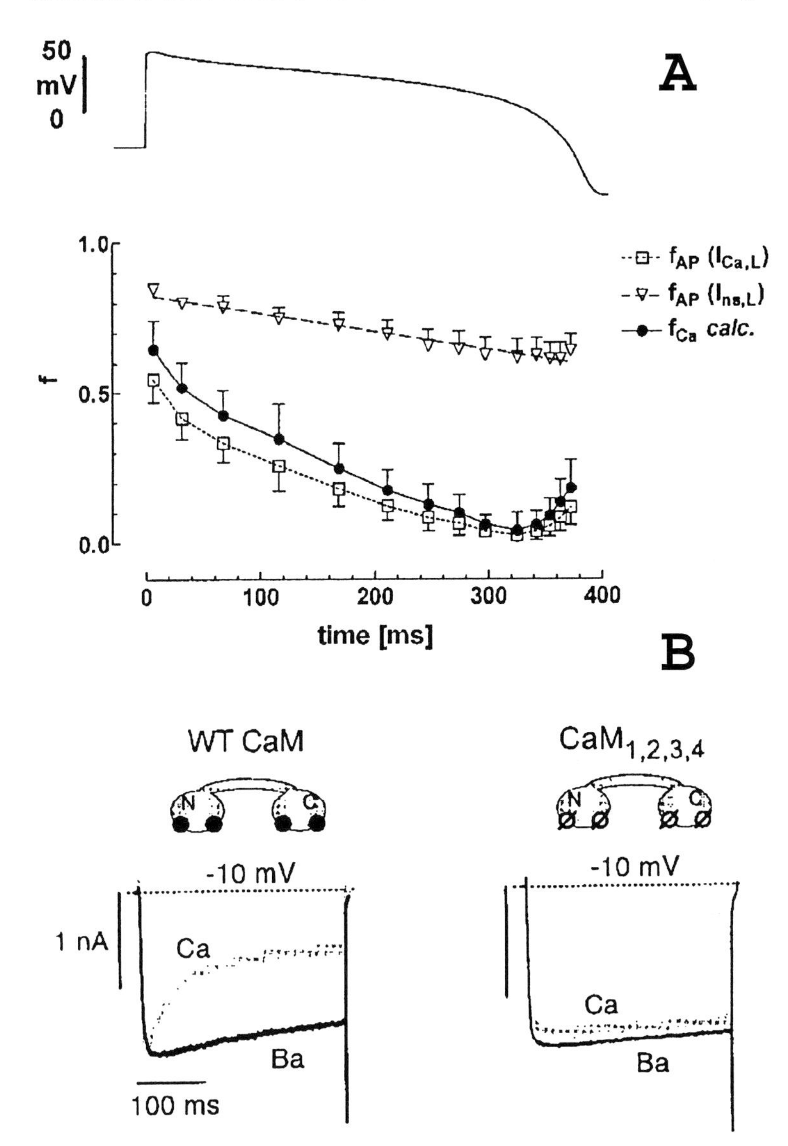

Fig. 3. (**A**) AP clamp waveform and estimates of the fraction of LCCs *not* voltage-inactivated (open triangles) and *not* Ca^{2+}-inactivated (*filled circles*) during the AP-clamp. Data measured in guinea pig ventricular myocytes. Reproduced from Fig. 11A–B of Linz and Meyer [12], with permission. (**B**) L-Type Ca^{2+} currents recorded from wild-type (WT CaM) rat ventricular myocytes or myocytes overexpressing mutant CaM ($CaM_{1,2,3,4}$). Responses are to voltage-clamp steps from a holding potential of -80 to -10 mV in the presence of Ca^{2+} (*dotted line*) or Ba^{2+} (*solid line*) as the charge carrier. Reproduced from Fig. 2 of Peterson et al. [10], with permission

of the channel and dissociation of FKBP12.6 increases RyR open probability and Ca^{2+} sensitivity [13]. Evidence suggests that cardiac RyRs may be physically coupled and exhibit coordinated gating which is functionally ablated in the absence of FKBP12.6 [14]. Cardiac RyRs contain a PKA phosphorylation site at serine-2809. The effects of PKA-induced phosphorylation of the RyR remains controversial, with one group asserting that phosphorylation leads to dissociation of FKBP12.6 from the RyR [15], thus regulating coordinated gating and channel open probability, while another group asserts that it does not [16]. Ca^{2+}/calmodulin-dependent protein kinase II (CaMKII) also phosphorylates RyR at serine-2815, leading to increased channel open probability and increased Ca^{2+} sensitivity [17].

2.2.3 Calcium-Induced Calcium-Release

EC Coupling involves a close interplay between LCCs and RyRs within the dyadic space. During the initial depolarization stages of the AP, voltage-gated LCCs open, allowing the entry of Ca^{2+} into the dyad. As Ca^{2+} concentration in the dyad increases, Ca^{2+} binds to the RyR, increasing their open probability and leading to Ca^{2+} release from the JSR – a process called Ca^{2+}-induced Ca^{2+} release (CICR).

The phenomenon of CICR has been studied extensively. Experiments have shown two major properties of CICR: (1) graded Ca^{2+} release; and (2) voltage-dependent EC coupling gain. *Graded release* refers to the phenomenon originally observed by Fabiato and co-workers [18–20] that Ca^{2+} release from JSR is a smooth and continuous function of trigger Ca^{2+} entering the cell via LCCs. Figure 4 (adapted from Fig. 3 of Wier et al. [21]) shows experimentally measured properties of graded release. Figure 4A shows trigger flux of Ca^{2+} through LCCs (open circles, F_{ICa}) and release flux of Ca^{2+} through RyRs (filled circles, $F_{\mathrm{SR,rel}}$) in response to a range of depolarizing voltage-clamp steps. Both are smooth continuous functions of step potential. Figure 4B shows normalized versions of the flux curves. The curve for RyR release flux is shifted in the hyperpolarizing direction with respect to that for LCC trigger flux. This occurs since at lower membrane potentials, single LCC currents are large and thus highly effective at triggering opening of RyR. At higher potentials, closer to the reversal potential for Ca^{2+}, open probability of LCCs is high, but single channel currents are small and thus less effective at triggering opening of the RyR. *EC coupling gain* is defined as the ratio of peak Ca^{2+} release flux through RyRs to the peak trigger flux through LCCs. There are more RyRs than LCCs in mammalian cardiac ventricular cells, with the RyR:LCC ratio varying from 8:1 in rat, 6:1 in humans to 4:1 in guinea pig [22]. The result is that a greater amount of Ca^{2+} is released from JSR via the RyR than enters the cell through LCCs, leading to high EC coupling gain. Figure 4C shows an example of the EC coupling gain function in rat ventricular myocytes. The voltage dependence of gain arises from the relative displacement of the RyR and LCC flux curves shown in Fig. 4B.

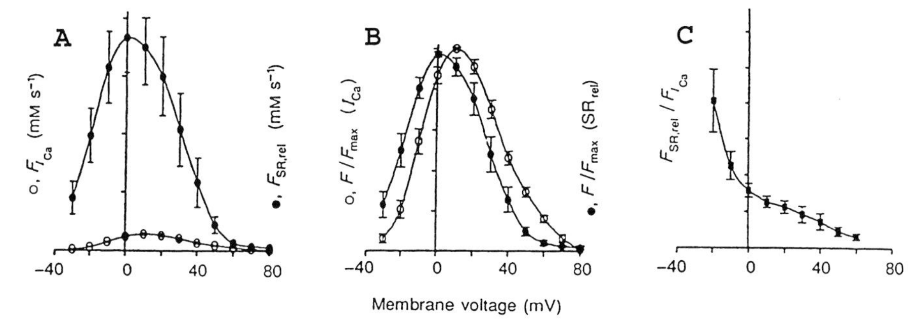

Fig. 4. (**A**) Ca^{2+} flux (ordinate, mM-s^{-1}) through RyRs (*filled circles*) and LCCs (*open circles*) as a function of membrane potential (abscissa, mV) measured from rat ventricular myoyctes. (**B**) Normalized data from Fig. 4A. (**C**) EC coupling gain (ordinate) as a function of membrane potential (abscissa, mV). Reproduced from Wier et al. [21] Fig. 3, with permission

2.2.4 Ca^{2+} Re-Uptake and Extrusion

Several mechanisms exist to restore Ca^{2+} concentration to resting levels following an AP (Fig. 2). These mechanisms are the Na^+-Ca^{2+} exchanger, the sarcolemmal Ca^{2+}-ATPase and the SR Ca^{2+}-ATPase [23]. The Na^+-Ca^{2+} exchanger is generally believed to import three Na^+ ions for every Ca^{2+} ion extruded, yielding a net charge movement [24]. More recent data support the idea that the exchange ratio may be 4:1 [25] or even variable [26]. The Na^+-Ca^{2+} exchanger is believed to be located predominantly in the sarcolemmal membrane and is driven by both transmembrane voltage and Na^+ and Ca^{2+} concentration gradients. It can work in forward mode, in which case it extrudes Ca^{2+} and imports Na^+, thus generating a net inward current, or in reverse mode, in which case it extrudes Na^+ and imports Ca^{2+} thus generating a net outward current. Some experimental evidence suggests that during the plateau phase of the AP, the Na^+-Ca^{2+} exchanger works initially in reverse mode bringing Ca^{2+} into the cell, and later switches to forward mode thereby extruding Ca^{2+} [27]. The second Ca^{2+} extrusion mechanism is the sarcolemmal Ca^{2+}-ATPase. This Ca^{2+} pump hydrolyzes ATP to transport Ca^{2+} out of the cell. It contributes a sarcolemmal current which is small relative to that of the Na^+-Ca^{2+} exchanger. At equilibrium, during each cardiac cycle (the time from the start of one AP to the next) the total amount of Ca^{2+} leaving the cell via the Na^+-Ca^{2+} exchanger and the sarcolemmal Ca^{2+}-ATPase is equal to the amount that enters. The third extrusion mechanism for myoplasmic Ca^{2+} is the SR Ca^{2+}-ATPase. This ATPase sequesters Ca^{2+} into the SR. The SR Ca^{2+}-ATPase is predominantly located in the NSR membrane. In equilibrium, during each cardiac cycle the SR Ca^{2+}-ATPase re-sequesters an amount of Ca^{2+} equal to that released by the SR via the RyR. The SR Ca^{2+}-ATPase hydrolyzes ATP to transport Ca^{2+}, and has both forward and reverse modes [28, 29]. The reverse mode serves to prevent overloading of the SR with Ca^{2+} at rest.

3 Computational Models of Cardiac EC Coupling

3.1 Common Pool Models of CICR

In the majority of models of the cardiac ventricular myocyte, CICR is described either phenomenologically or through use of a formulation known as the "common pool" model. As defined by Stern [30] and as illustrated in Fig. 5, common pool models [31–33] are ones in which Ca^{2+} flux through all LCCs are lumped into a single trigger flux, Ca^{2+} flux through all RyRs is lumped into a single release flux and both the trigger and release flux are directed into a common Ca^{2+} compartment (labeled the "subspace" in Fig. 5). Consequently, in such models, activation of the JSR release mechanism is controlled by Ca^{2+} concentration in this common pool. The result of this physical arrangement is that once RyR Ca^{2+} release is initiated, the resulting increase of Ca^{2+} concentration in the common pool stimulates regenerative, all-or-none rather than graded Ca^{2+} release [30]. This "latch up" of Ca^{2+} release can be avoided and graded JSR release can be achieved in phenomenological models of EC coupling by formulating Ca^{2+} release flux as an explicit function of only sarcolemmal Ca^{2+} flux rather than as a function of Ca^{2+} concentration in the common pool [34–37]. Models of this type are not common pool models based on the definition given by Stern, and do not suffer an inability to exhibit both high gain and graded JSR Ca^{2+} release. However, such phenomenological formulations lack mechanistic descriptions of the processes that are the

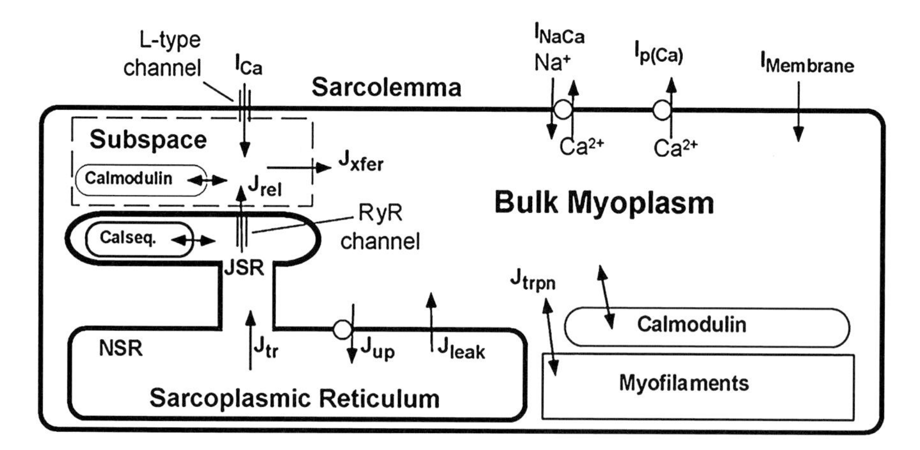

Fig. 5. Structure of a common pool myocyte model. Reproduced from Jafri et al. [31] Fig. 1, with permission

underlying basis of CICR. We therefore focus, in the remainder of this presentation, on a review of the strengths and weaknesses of the common pool model formulation.

3.1.1 Strengths of Common Pool Models

Despite the inability of common pool models to describe the fundamental property of graded JSR Ca^{2+} release, such models, when incorporated into integrative models of the myocyte, have proven reconstructive and predictive abilities. As will be discussed subsequently (Sect. 4.1), myocyte models based on the common pool formulation have been able to reconstruct many aspects of the cellular phenotype of ventricular myocytes isolated from end-stage failing canine and human hearts including prolongation of AP duration, reduced amplitude of intracellular Ca^{2+} transients and slowed decay of the Ca^{2+} transient (see Fig. 1 of Winslow et al. [32]). A common pool model of the guinea pig left ventricular myocyte has been able to reconstruct properties of extrasystolic restitution and post-extrasystolic potentiation in response to so-called S1-S2 stimulus protocols (see Fig. 5B of Rice et al. [38]). A recent computational model of the human left ventricular myocyte is able to predict rate-dependent prolongation of AP duration, AP duration restitution curves and Ca^{2+}-frequency relationships (see Figs. 5A, 6A and 7C of Iyer et al., respectively [39]). Thus, such models are able to account for a broad range of responses measured experimentally.

3.1.2 Weaknesses of Common Pool Models

In light of the successes detailed above, one may question whether graded JSR Ca^{2+} release plays an important role in shaping the electrophysiological responses of the ventricular myocyte. The answer is a definite “yes” and the reason has to do with recent data concerning the relative contribution of Ca^{2+}- versus voltage-dependent inactivation of LCCs reviewed in Sect. 2.2.1 and shown in Fig. 3. These data demonstrate that LCC Ca^{2+}-dependent inactivation is strong with rapid onset, whereas voltage-dependent inactivation is weaker with slower onset. In contrast, Fig. 6 shows the relationship between Ca^{2+}- and voltage-dependent LCC inactivation in three different models of the cardiac AP. Figures 6A–B replicate the experimental data of Linz and Meyer [12] and Figs. 6C–E show APs and Ca^{2+}- versus voltage-dependent inactivation of LCCs during these APs for common pool canine [32] and guinea pig [31] myocyte models and the Luo-Rudy guinea pig model incorporating a phenomenological description of CICR [36]. For each model, Ca^{2+} inactivation of LCCs is smaller or comparable in magnitude to voltage-dependent inactivation. This stands in contrast to the experimental data shown in Fig. 3.

The consequence of incorporating the relationship between Ca^{2+}- versus voltage-dependent inactivation shown in Fig. 3 into a common pool model

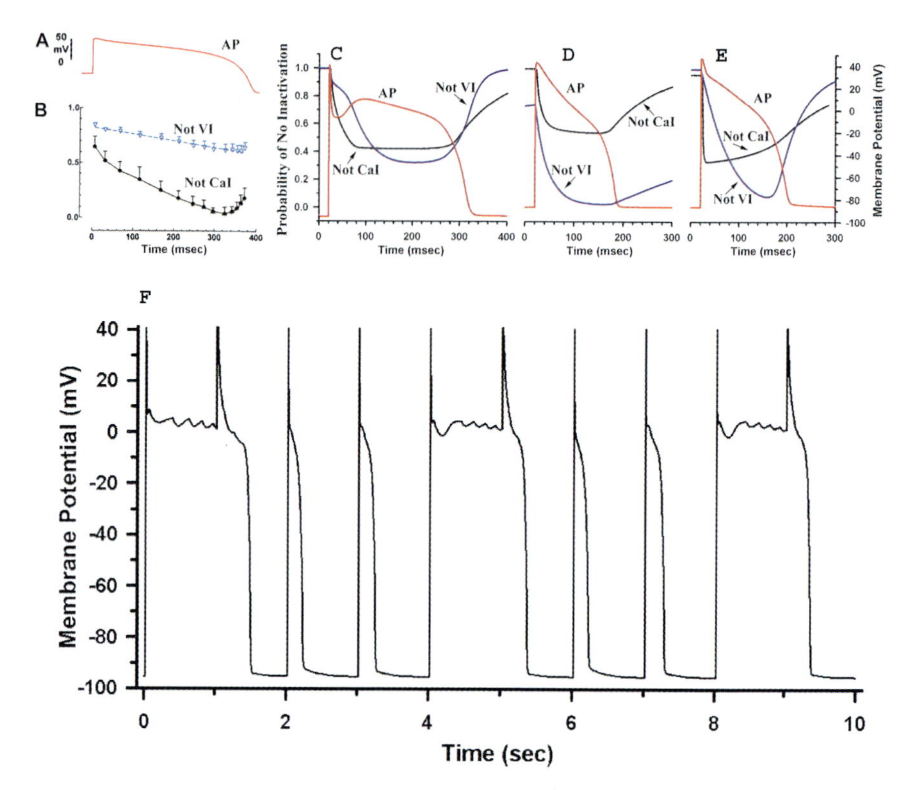

Fig. 6. Action potentials, fraction of LCCs not Ca^{2+}-inactivated (Not CaI) and not voltage-inactivated (Not VI) during the AP. (**A**) AP-clamp waveform used to control membrane potential of myocytes isolated from guinea pig left ventricle. (**B**) Experimental estimates of the fraction of LCCs not Ca^{2+}-inactivated and not voltage-inactivated when membrane potential is clamped as shown in Panel A. Panels C-E show similar estimates based on simulated actions potentials. Results are shown for APs generated using the Winslow et al. [32] (**C** – canine), Jafri-Rice-Winslow [31] (**D** – guinea pig), and Luo-Rudy Phase II [36] (**E** – guinea pig) ventricular myocyte models. **F**) Membrane potential as a function of time for a 10-second simulation of a modified version of the Winslow et al. [32] model with I_{CaL} parameterized with strongly Ca^{2+}-dependent and weakly voltage-dependent inactivation (similar to that of the local control model). Panels A and B are reproduced from Fig. 11 of Linz and Meyer [12], with permission

of the canine ventricular myocyte [32] is shown in Fig. 6F. AP duration becomes unstable. The reason for this is that in a model where the release of JSR Ca^{2+} is controlled by sensing Ca^{2+} levels in the same pool into which Ca^{2+} is released, Ca^{2+} release occurs in an all-or-none fashion [30]. When Ca^{2+}-dependent inactivation of LCCs is the dominant inactivation process in such common pool models, LCC inactivation also exhibits all-or-none behavior, switching on in response to JSR Ca^{2+} release and switching off in its absence. This switching between all-or-none LCC inactivation destabilizes the plateau phase of the AP, with APs alternating between those with short (~150–250 ms) and long (>1000 ms) duration. This unstable behavior occurs over a wide range of LCC inactivation parameters as long as voltage-dependent inactivation of LCCs is relatively slow and weak. Strong voltage-dependent inactivation of I_{CaL}, although contrary to experimental observations, is therefore necessary to enforce stability of common pool models. When new data regarding the balance between voltage- and Ca^{2+}-dependent inactivation is incorporated into these models, they fail at reproducing even the most elementary electrophysiological response of the ventricular myocyte – a stable AP.

3.2 A Stochastic Local-Control Model of CICR

The fundamental failure of common pool models described above suggests that more biophysically-based models of EC coupling must be developed and investigated. Understanding of the mechanisms by which Ca^{2+} influx via LCCs triggers Ca^{2+} release from the JSR has advanced tremendously with the development of experimental techniques for simultaneous measurement of LCC currents and Ca^{2+} transients and detection of local Ca^{2+} transients and this has given rise to the local control hypothesis of EC coupling [21, 30, 40, 41]. This hypothesis asserts that opening of an individual LCC in the T-tubular membrane triggers Ca^{2+}-release from the small cluster of RyRs located in the closely apposed (~10 nm) JSR membrane. Thus, the local control hypothesis asserts that release is all-or-none at the level of these individual groupings of LCCs and RyRs. However, LCC and RyR clusters are physically separated at the ends of the sarcomeres [42]. These clusters therefore function in an approximately independent fashion. The local control hypothesis asserts that graded control of JSR Ca^{2+} release, in which Ca^{2+}-release from JSR is a smooth, continuous function of Ca^{2+} influx through LCCs, is achieved by the statistical recruitment of elementary Ca^{2+} release events in these independent dyadic spaces. Thus, at the heart of the local-control hypothesis is the assertion that the co-localization of LCCs and RyRs is a structural component that is fundamental to the property of graded Ca^{2+} release and force generation at the level of the cell.

Several computational models have been developed to investigate properties of local Ca^{2+} release at the level of the cardiac dyad [5, 43–45]. Each of these model formulations incorporates: (1) one or a few LCCs; (2) a cluster of RyRs; (3) the dyadic volume in which the events of CICR occur; and (4) anionic binding sites which buffer Ca^{2+}. In some of these models, detailed descriptions of diffusion and Ca^{2+} binding in the dyadic cleft are employed to demonstrate the effects of geometry, LCC and RyR properties and organization, and surface charge on the spatio-temporal profile of Ca^{2+} within the dyad, and hence on the efficiency of CICR [5, 44, 45]. Stern et al. [46] have simulated CICR stochastically using numerous RyR schemes to demonstrate conditions necessary for stability of EC coupling and have suggested a possible role for allosteric interactions between RyRs. The functional release unit model of Rice et al. [43] has demonstrated that local control of CICR can be obtained without including computationally intensive descriptions of Ca^{2+} gradients within the dyadic space. Isolated EC coupling models such as these, however, cannot elucidate the nature of the interaction between local events of CICR and integrative cellular behavior, as the models are simply too computationally demanding.

We have recently developed a local-control model of CICR which is sufficiently minimal that it may be incorporated within a computational model of the cardiac ventricular myocyte [47]. This model is derived from a canine ventricular myocyte model [32] and incorporates stochastic gating of LCCs and RyRs. This model has been shown to capture fundamental properties of local control of CICR such as high gain, graded release and stable release termination. The model incorporates: (1) sarcolemmal ion currents of the Winslow et al. canine ventricular cell model [32]; (2) continuous time Markov chain models of the rapidly-activating delayed rectifier potassium current I_{Kr} [48], the Ca^{2+}-independent transient outward K current I_{to1} [49] and the Ca^{2+}-dependent transient outward chloride (Cl^-) current I_{to2}; (3) a continuous-time Markov chain model of I_{CaL} in which Ca^{2+}-mediated inactivation occurs via the mechanism of mode-switching [31]; (5) an RyR channel model adapted from that of Keizer and Smith [50]; and (6) locally controlled CICR from junctional sarcoplasmic reticulum (JSR) via inclusion of LCCs, RyRs, chloride channels, local JSR and dyadic subspace compartments within Ca^{2+} release units (CaRUs).

Figure 7 shows a schematic of the CaRU. The CaRU model is intended to mimic the properties of Ca^{2+} sparks in the T-tubule/SR (T-SR) junction. Figure 7B shows a cross-section of the model T-SR cleft, which is divided into four individual dyadic subspace compartments arranged on a 2×2 grid. Each subspace (SS) compartment contains a single LCC and 5 RyRs in its JSR and sarcolemmal membranes, respectively. All 20 RyRs in the CaRU communicate with a single local JSR volume. The 5:1 RyR to LCC stoichiometry is consistent with recent estimates indicating that a single LCC typically triggers the opening of 4–6 RyRs [51]. Each subspace is treated as a single compartment in which Ca^{2+} concentration is uniform, however Ca^{2+} may diffuse passively to neighboring subspaces within the same CaRU. The division of the CaRU into four subunits allows for the possibility that an LCC may trigger Ca^{2+} release in adjacent subspaces (i.e., RyR recruitment) under conditions where unitary LCC currents are large. The existence of communication among adjacent subspace volumes is supported by the findings that Ca^{2+} release sites can be coherent over distances larger than that occupied by a single release site [52], and that the mean amplitude of Ca^{2+} spikes (local SR Ca^{2+} release events that consist of one or a few Ca^{2+} sparks [53]), exhibits a bell shaped voltage dependence, indicating synchronization of multiple Ca^{2+} release events within

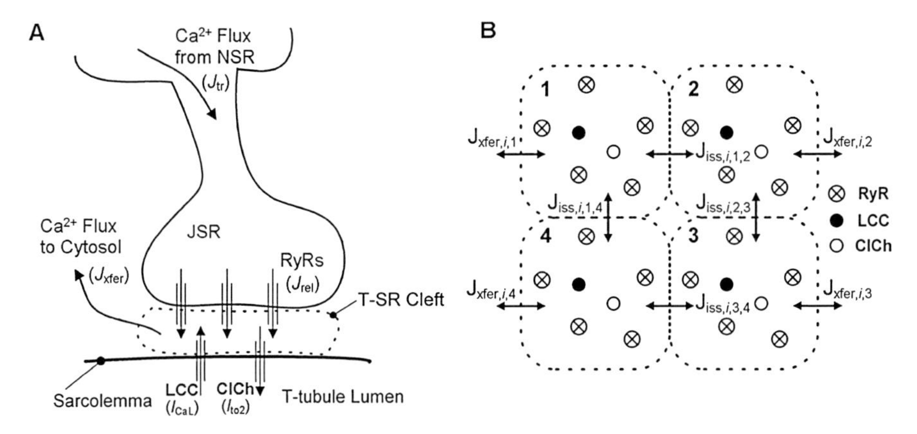

Fig. 7. Schematic representation of the Ca^{2+} release unit model (CaRU). (**A**) Trigger Ca^{2+} influx through the LCCs enters into the T-SR cleft (dyadic space). The rise in local Ca^{2+} level promotes the opening of RyRs and Ca^{2+}-modulated chloride channels (ClChs). The excess local Ca^{2+} passively diffuses out of the cleft into the cytosol and JSR Ca^{2+} is refilled via passive diffusion from the NSR. (**B**) The T-SR cleft (shown in cross-section) is composed of four dyadic subspace volumes, arranged on a 2×2 grid, each containing 1 LCC, 1 ClCh, and 5 RyRs. Ca^{2+} in any subspace may diffuse to a neighboring subspace (J_{iss}) or to the cytosol (J_{xfer}). $J_{iss,i,j,l}$ represents Ca^{2+} flux from the jth subspace to the lth subspace within the ith CaRU. Similarly $J_{xfer,i,j}$ represents Ca^{2+} flux from the jth subspace to the cytosol from the ith CaRU

a T-SR junction [54]. The choice of four subunits allows for semi-quantitative description of dyadic Ca^{2+} diffusion while limiting model complexity.

L-type Ca^{2+} current (I_{CaL}) is a function of the total number of channels (N_{LCC}), single channel current magnitude (i), open probability (p_o), and the fraction of channels that are available for activation (f_{active}), where $I_{CaL} = N_{\mathrm{LCC}} \times f_{\mathrm{active}} \times i \times p_o$. Single LCC parameters are based on experimental constraints on both i and p_o. The product $N_{\mathrm{LCC}} \times f_{\mathrm{active}}$ is chosen such that the amplitude of the whole-cell current agrees with that measured experimentally in canine myocytes. This approach yields a value of 50,000 for $N_{\mathrm{LCC}} \times f_{\mathrm{active}}$, consistent with experimental estimates of active LCC density and corresponding to 12,500 active CaRUs.

One of the bases for local control of SR Ca^{2+} release is the structural separation of T-SR clefts at the ends of sarcomeres (i.e., RyR clusters are physically separated) [42]. Each CaRU is therefore simulated independently in accord with this observation. Upon activation of RyRs, subspace Ca^{2+} concentration increases. This Ca^{2+} will diffuse freely to either adjacent subspace compartments (J_{iss}) or into the cytosol (J_{xfer}) as determined by local concentration gradients. The local JSR compartment is refilled via passive diffusion of Ca^{2+} from the network SR (NSR) compartment (J_{tr}).

The algorithm for solving the stochastic ordinary differential equations defining the model has been described previously [47]. Briefly, transition rates for each channel are determined by their gating schemes and their dependence on local Ca^{2+} level. Gating of each channel within a CaRU is simulated by choosing channel occupancy time as an exponentially distributed random variable with parameter determined by the sum of voltage- and/or Ca^{2+}-dependent transition rates from the current state. Stochastic simulation of CaRU dynamics is used to determine all Ca^{2+} flux into and out of each local subspace. The summation of all Ca^{2+} fluxes crossing the CaRU boundaries are taken as inputs to the global model, which is defined by a system of coupled ordinary differential equations. The dynamical equations defining the global model are solved using the Merson modified Runge-Kutta 4th-order adaptive time step algorithm which has been modified to embed the stochastic CaRU simulations within each time step.

This model provides the ability to investigate the ways in which LCC, RyR and subspace properties impact on CICR and the integrative behavior of the myocyte. However, this ability is achieved at a high computational cost. Three approaches are therefore used to accelerate the computations. First, the pseudo-random number generator used is the Mersenne Twister algorithm [55], having a long period (2^{19937}–1) and reduced computation time. Second, we have developed an algorithm for the dynamic allocation of model CaRUs so that a large number of CaRUs are utilized prior to, during and shortly after the AP, and a smaller number of CaRUs is used during diastole. Third, simulation of the stochastic dynamics of the independent functional units may be performed in parallel, resulting in near linear speedup as the number of processors is increased. These modifications enable us to simulate up to 1 Sec of model activity in 1 minute of simulation time when running on 6 IBM Power4 processors configured with 4 Gbytes memory each.

Figure 8 shows macroscopic properties of APs and SR Ca^{2+} release in this hybrid stochastic/ODE model. Figure 8A shows the voltage dependence of peak Ca^{2+} flux (ordinate) through LCCs (filled circles) and RyRs (open circles) as a function of membrane potential (mV, abscissa). Unlike the case for common pool models, Ca^{2+} release flux is a smooth and continuous function of membrane potential, and hence trigger Ca^{2+}, as shown by the experimental data in Fig. 4A. Figure 8B shows the data of Fig. 8A following normalization. The model data exhibit the hyperpolarizing shift of release flux relative to trigger flux seen in the experimental data of Fig. 4B. EC coupling gain (ordinate) is shown as a function of membrane potential (mV, abscissa) in Fig. 8C. Open boxes show gain when there is no Ca^{2+} diffusion between the 4 functional release units comprising each CaRU, and the open triangles show gain when such diffusion is accounted for. In the presence of Ca^{2+} diffusion between functional units comprising each CaRU, EC coupling gain is greater at all potentials, but the increase in gain is most dramatic at more negative potentials. In this negative voltage range, LCC open probability is sub-maximal, leading to sparse LCC openings. However, unitary current magnitude is

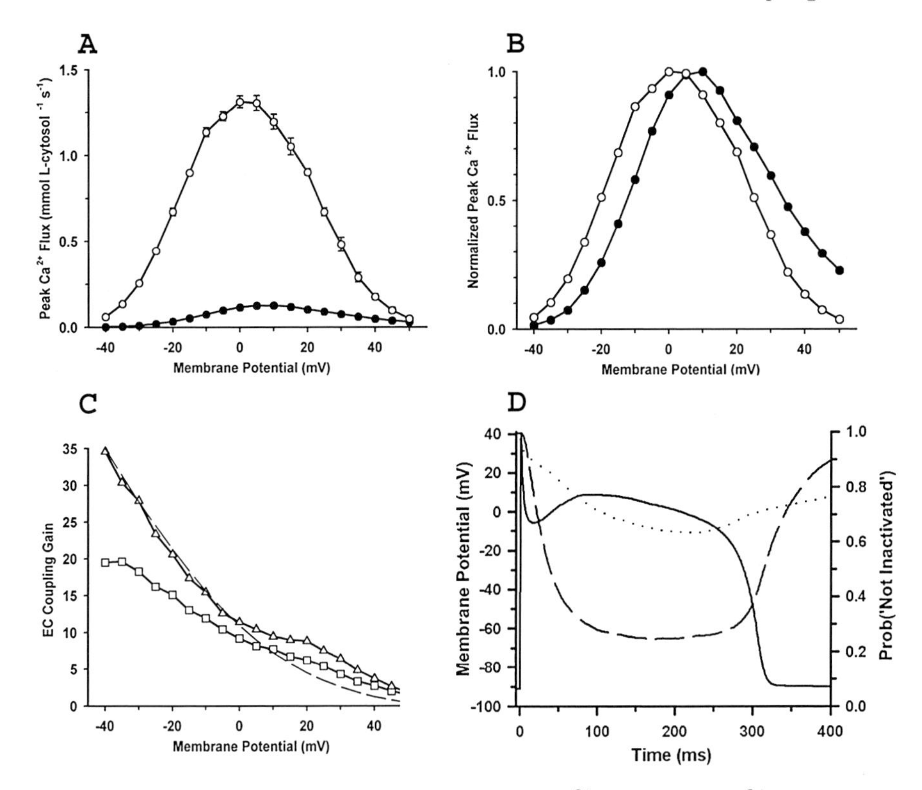

Fig. 8. Voltage dependence of macroscopic LCC Ca^{2+} influx, SR Ca^{2+} release, and EC coupling gain. (**A**) Mean peak Ca^{2+} flux amplitudes, $F_{LCC(max)}$ (*filled circles*) and $F_{RyR(max)}$ (*open circles*) as a function of membrane voltage, $n = 5$ simulations at each voltage. (**B**) Peak Ca^{2+} fluxes (data of panel A) normalized by their respective maxima. (**C**) EC coupling gain as a function of membrane potential defined as $F_{RyR(max)}/F_{LCC(max)}$ under control conditions (*triangles*) and in the absence of intersubspace coupling within the CaRUs (*squares*), as well as L-type unitary current (*dashed line*, scaled to match the gain function at –40 mV). (**D**) Action potential (*solid line*) obtained in the stochastic local control model incorporating the relationship between V- and Ca^{2+}-dependent inactivation measured by Linz & Meyer [12]

relatively high, so that in the presence of Ca^{2+} diffusion within the CaRU, the rise in local Ca^{2+} due to the triggering action of a single LCC can recruit and activate RyRs in adjacent subspace compartments within the same T-SR junction. The net effect of inter-subspace coupling is therefore to increase the magnitude and slope of the gain function preferentially in the negative voltage range.

Figure 8D shows the relative balance between the fraction of LCCs *not* voltage-inactivated (dotted line) and *not* Ca^{2+}-inactivated (dashed line) during an AP simulated using the local-control ventricular myoycte model. These fractions were designed to fit the experimental data of shown in Fig. 3A. The solid line shows a local-control model AP. This AP should be contrasted with those produced by the common pool model (Fig. 6F) when the same relationship between LCC voltage- and Ca^{2+}-dependent inactivation is used. Clearly, the local-control model exhibits stable APs whereas the common pool model does not. These simulations therefore offer an intriguing glimpse of the functional importance of local control of CICR in shaping properties of the AP and of how co-localization and stochastic gating of individual channel complexes can have a profound effect on the integrative behavior of the myocyte.

3.3 Coupled LCC-RyR Gating Models of CICR

The results described above demonstrate that to accommodate new data regarding strong negative feedback regulation of LCC function by JSR Ca^{2+} release, myocyte models must incorporate graded CICR. Unfortunately, local control models based on stochastic simulation of CaRU dynamics remain far too computationally demanding to be used routinely by most laboratories in single cell simulations, let alone in models of cardiac tissue.

To address this problem, we have recently formulated a novel model of CICR which describes the underlying channels and local control of Ca^{2+} release, but consists of a low dimensional system of ordinary differential equations [56]. This is achieved in two steps, using the same techniques as applied by Hinch in an analysis of the generation of spontaneous sparks in a model of a cluster of RyRs [57]. First, the underlying channel and CaRU models are minimal, such that they only contain descriptions of the essential biophysical features observed in EC coupling. This in turn yields a system of model equations which can be simplified by applying approximations based on a separation of time-scales. In particular, it can be shown that Ca^{2+} in the dyadic space equilibrates rapidly relative to the gating dynamics of LCCs and RyRs. The joint behavior of LCCs and RyRs can then be described using a Markov model where the transition probabilities between interacting states are a function of global variables only. This in turn allows the ensemble behavior of the CaRUs to be calculated using ordinary differential equations. The resulting model, which we refer to as the coupled LCC-RyR gating model, has parameters which may be calculated directly from the underlying biophysical model of local control of Ca^{2+} release. Despite the simplicity of this model, it captures key properties of CICR including graded release and voltage-dependence of EC coupling gain. The model is therefore well suited for incorporation within single cell and tissue models of ventricular myocardium.

3.3.1 A "Minimal" Coupled LCC-RyR Gating Model

The model of local-control of CICR which we will develop is minimal in the sense that: (a) CaRUs consist of only one LCC, one RyR and the subspace within which they communicate; and (b) simplified continuous time Markov models of LCC and RyR gating are employed, each consisting of three-states.

Despite this simplicity, the model is able to describe key properties of local-control of CICR. The following sections provide a brief overview of model development. Full details are given in Hinch et al. [57].

As a starting point for development of a minimal LCC model, we use the LCC model developed by Jafri et al. [31] and modified by Greenstein and Winslow [47]. This model was formulated based on the molecular structure of the channel which is assumed to be composed of four independently gating subunits [31]. This leads to an activation process described by five closed states. The model is shown in Fig. 9A. LCCs are assumed to gate in a “Mode Normal” in which they are not Ca^{2+}-inactivated, and a “Mode Ca” in which they are Ca^{2+}-inactivated. Horizontal transitions in either mode are voltage-dependent, with rightward transitions corresponding to channel activation following membrane depolarization. In Mode Normal, the final transition from state C_4 to the open conducting state O is voltage-independent. In Mode Ca, transitions from IC_4 to an open state do not exist, corresponding to channel inactivation. Vertical transitions are voltage-independent, with transition rates from Mode Normal to Mode Ca being a function of Ca^{2+} concentration in the dyadic space, denoted as $[Ca^{2+}]_{ds}$.

The model is first simplified by reducing the number of closed and closed-inactivated states. The resulting model is shown in Fig. 9B. The model consist of two closed states (denoted C_3 and C_4), a single open state O accessible from closed state C_4 and two Ca^{2+}-inactivated states IC_3 and IC_4. While the structure of this reduced 5-state model is no longer based on the molecular structure of the LCC, it retains the essential functional features of the full model such as well-defined gating modes and rates of Ca^{2+}-mediated inactivation which depend upon activation (rightward transitions) of the channel. This model can be simplified further to a 3-state model (Fig. 9C) by using the fact that transitions between the state pairs C_3 and C_4 and IC_3 and IC_4 are rapid relative to the transition rates between these two sets of states. Define the combined closed state $C = C_3 \cup C_4$ and the combined inactivated state $I = IC_3 \cup IC_4$. Since the time-scale of the transitions between C_3 and C_4 is the smallest time-scale in the model, we can assume that these 2 states are in equilibrium and thus define conditional state occupancy probabilities as

$$P(C_3/C) = \frac{a_{-1}}{a_{-1} + a_1}$$

$$P(C_4/C) = \frac{a_1}{a_1 + a_{-1}}$$

A similar approximation is applied to states IC_3 and IC_4. Under these assumptions, the forward transition rate between the combined closed state C and the combined inactivated state I is given by

$$\varepsilon_+([\mathrm{Ca}^{2+}]_{\mathrm{ds}}) = \mathrm{a}\varepsilon_1([\mathrm{Ca}^{2+}]_{\mathrm{ds}})\mathrm{P}(C_3|C) + \varepsilon_1([\mathrm{Ca}^{2+}]_{\mathrm{ds}})\mathrm{P}(C_4|C)$$

A similar approach may be used to derive the remaining transition rates between states I, C and O. The open channel current is then given by the

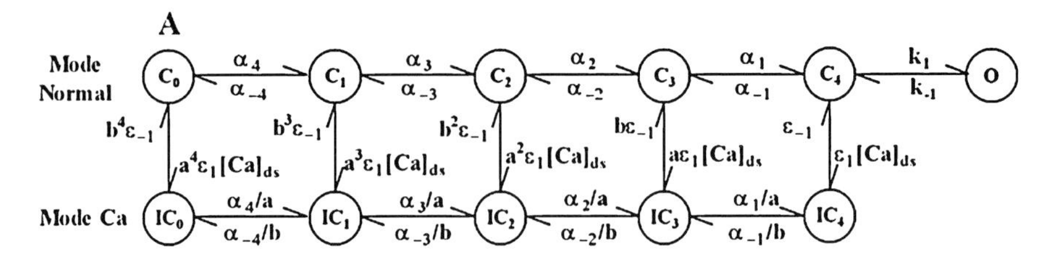

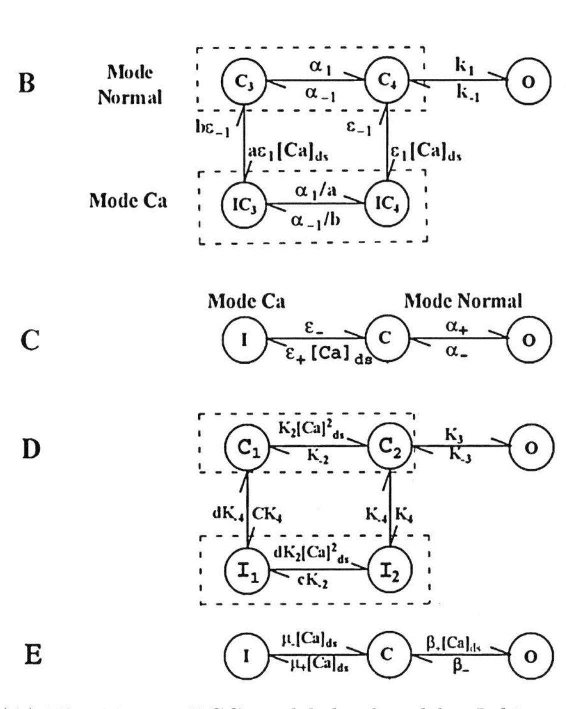

Fig. 9. (**A**) The 11-state LCC model developed by Jafri et al. [31]. (**B**) Simplification of the model in (**A**) by truncation of the leftmost 3 columns of states. (**C**) Simplification of the model in (**B**) by applying the rapid equilibrium approximation to state pairs C_3 and C_4 and IC_3 and IC_4. (**D**) 5-state RyR model from Stern et al. [46]. (**E**) Simplification of the 5-state model in (**D**) by application of the rapid equilibrium approximation to state pairs C_1 and C_2 and I_1 and I_2

Goldman-Hodgkin-Katz equation (see Hinch et al. [56]). This model of the LCC will be validated against experimental data in the following section.

We use a 5-state model of the RyR (Fig. 9D) based on Scheme 6 of Stern et al. [46], with the addition of modal gating between states C and O. Transitions from the closed to open modes occur upon binding of two Ca^{2+} ions. Transitions between states C_1 and C_2, and inactivated states I_1 and I_2 are assumed to be rapid. Following the same procedure used in the reduction of the LCC model, the RyR model is reduced to a 3-state model (Fig. 9E).

The magnitude of the Ca^{2+} flux through an open RyR is proportional to the difference in Ca^{2+} concentration between the SR and the local dyadic space.

We employ a minimal model for each CaRU consisting of one LCC, one RyR and the dyadic space within which these channels reside. Experimental recordings of triggered Ca^{2+}-sparks show that a single LCC opening may activate 4–6 RyRs [51]. The CaRU model employed here is therefore a simplification of actual dyadic structure and function as described previously. Several prior models of CaRUs [47, 58] also include a local JSR volume which is depleted relative to the network SR during Ca^{2+} release. However, recent experimental studies suggest that the JSR Ca^{2+} is in quasi-equilibrium with network SR during Ca^{2+} release [59]. Therefore, the minimal CaRU model does not include a local JSR. Calcium flux from the dyadic space to the cytoplasm is governed by simple diffusion, such that the time-evolution of $[Ca^{2+}]_{ds}$ is given by

$$V_{ds}\frac{d[Ca^{2+}]_{ds}}{dt} = J_{RyR} + J_{LCC} - g_D([Ca^{2+}]_{ds} - [Ca^{2+}]_i)$$

where V_{ds} is the volume of the dyadic space, $[Ca^{2+}]_i$ is the Ca^{2+} concentration in the myoplasm, g_D is the conductance, and J_{RyR} and J_{LCC} are the currents through the RyR and LCC, respectively. The time constant of equilibrium of $[Ca^{2+}]_{ds}$, is given by $\tau_{ds} = V_{ds}/g_D \approx 3\,\mu s$ [57, 58]. Since this time constant is considerably smaller than that for opening of either LCC or RyR channels, we may use the rapid equilibrium approximation [57] to show that

$$[Ca^{2+}]_{ds} \sim [Ca^{2+}]_i + \frac{J_{RyR} + J_{LCC}}{g_D}$$

This is *the crucial step* in model simplification since $[Ca^{2+}]_{ds}$ is now a function of only the global variables $[Ca^{2+}]_{SR}$, $[Ca^{2+}]_i$, V and the state of the local RyR and LCC. As a result of this simplification, it is no longer necessary to solve a differential equation for each $[Ca^{2+}]_{ds}$ when modeling all CaRUs in the myocyte. Rather, $[Ca^{2+}]_{ds}$ is an algebraic function of the fluxes into and out of the dyadic space.

Armed with this simplification, it is now possible to define a state model in which each state describes the joint behavior of the LCC and RyR in each CaRU. Define Y_{ij} (where i, j = C, O, I) as the state of the CaRU with the LCC in the ith state and the RyR in the jth state. The CaRU can then be in one of 9 macroscopic states (Fig. 10). $[Ca]_{ds}$, J_{RyR} and J_{LCC} must be calculated separately for each of the 9 states. For example, consider the state Y_{co} with the LCC closed ($J_{LCC} = 0$) and the RyR open ($J_{RyR} = J_R\left([Ca^{2+}]_{SR} - [Ca^{2+}]_{ds}\right)$), then the rapid equilibrium approximation yields

$$c_{CO} = \frac{[Ca^{2+}]_i + \frac{J_R}{g_D}[Ca^{2+}]_{SR}}{1 + \frac{J_R}{g_D}}$$

$$J_{R,CO} = J_R \frac{[\mathrm{Ca}^{2+}]_{\mathrm{SR}} - [\mathrm{Ca}^{2+}]_i}{1 + \frac{J_R}{g_D}}$$

where $[\mathrm{Ca}^{2+}]_{\mathrm{ds}}^{\mathrm{co}}$ is $[\mathrm{Ca}^{2+}]_{\mathrm{ds}}$ in the state Y_{co} and $J_{\mathrm{RyR}}^{\mathrm{co}}$ is J_{RyR} in the state Y_{co}. Results for other states may be derived similarly. The resulting 9 state model of the CaRU shown in Fig. 10 is what we refer to as the coupled LCC-RyR gating model. Note that transitions are a function of $[\mathrm{Ca}^{2+}]_{\mathrm{ds}}$ which is itself a function of model states.

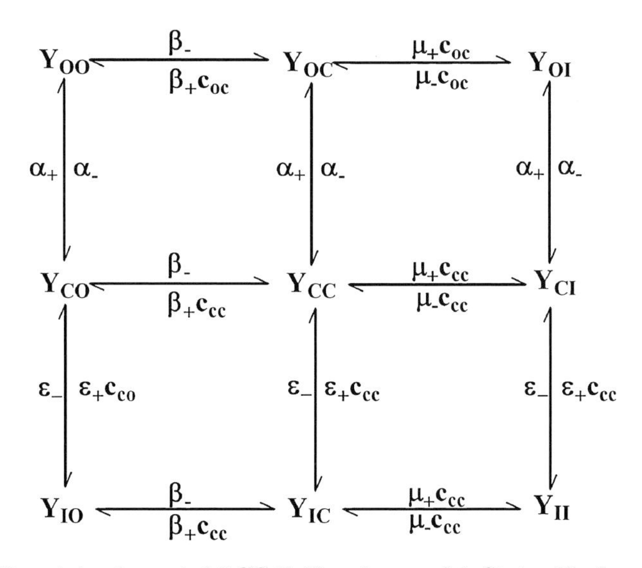

Fig. 10. The minimal coupled LCC-RyR gating model. States Y_{ij} denote the fraction of CaRUs in which the LCC is in state i and the RyR is in state j

The laws of mass action may next be used to derive a system of (eight) differential equations describing the time evolution of the fraction of LCC-RyR channels occupying the various states shown in Fig. 10. Whole cell Ca^{2+} currents are calculated by summing the contributions from the populations of CaRUs for which at least one LCC or RyR is open.

Figure 11A shows normalized peak flux through LCCs and RyRs (ordinate) as a function of membrane potential. As described previously, Ca^{2+} release is most effective at those membrane potentials producing large single LCC currents [21]. This results in the peak of the normalized JSR Ca^{2+} release flux being shifted by about 10 mV in the hyperpolarizing direction relative to the peak of the LCC trigger flux, as shown in the experimental data of Fig. 4B and in results from the stochastic local control model of Fig. 8B. This important behavior is also captured by the coupled LCC-RyR gating. As a consequence of this relative displacement of peak values, EC coupling gain decreases with increasing membrane potential. Figure 11B shows EC coupling

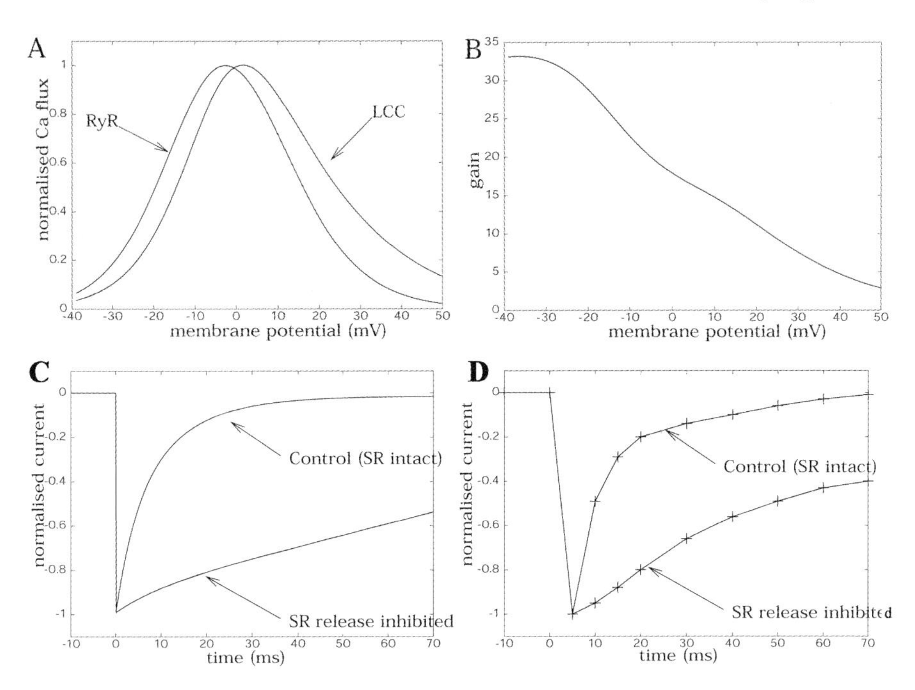

Fig. 11. (**A**) Normalized peak LCC and RyR flux (ordinate) as a function of membrane potential (abscissa, mV) for the minimal coupled LCC-RyR gating model. (**B**) EC coupling gain (ordinate) as a function of membrane potential (abscissa, mV) for the minimal coupled LCC-RyR gating model. Time course of LCC current after a voltage step from -50 mV to $+10$ mV for the minimal LCC-RyR coupled gating model (**C**) versus that measured experimentally in rat ventricular myocytes (**D**) [60]

gain (ordinate) as a function of membrane potential (abscissa, mV) predicted using the coupled LCC-RyR gating model. Results are in good agreement with both experimental data (Fig. 4C) and the stochastic local control model (Fig. 8C), thus demonstrating that despite its simplicity, the coupled LCC-RyR gating model is able to reconstruct the most critical feature of local control of CICR.

In voltage-clamp experiments in which JSR Ca^{2+} release is intact, it is found that the L-Type Ca^{2+} current is inactivated after approximately 20 ms. However, when JSR Ca^{2+} is depleted by application of caffeine, ryanodine or application of pre-pulses, Ca^{2+} inactivation is much slower. Figure 11C shows a comparison of the model prediction of this effect with experimental results (Fig. 11D) obtained in voltage-clamp studies using ventricular myocytes isolated from rat heart [60]. In both model and experiment, the cell was clamped at -50 mV and then stepped to 10 mV for 70 ms. JSR Ca^{2+} was depleted in the model by setting it to 10% of its normal value. Experimental and model results are in close agreement.

3.3.2 Generalized Coupled LCC-RyR Gating Models

The minimal nature of the model described above places some limits on its suitability for EC coupling and whole myocyte simulation studies. Such a model will be appropriate for large scale, multi-cellular simulations, but may be insufficient to quantitatively predict more complex cellular dynamics which depend upon the details of CICR. The method described above for deriving a coupled LCC-RyR, however, is not limited only to such highly reduced minimal models. We have developed a technique to build a coupled LCC-RyR gating model based on a CaRU which may contain one or more LCCs and/or RyRs, and where the individual channel models may maintain a greater level of complexity. Manual derivation of the equations describing such models would be an inefficient and error prone process because increased complexity and/or number of individual channel models leads to rapid growth in the number of equations required to describe the coupled system. For example, a CaRU with one 10-state LCC and five 4-state RyRs (see Fig. 12) will become a coupled LCC-RyR model consisting of 560 states. We have therefore designed and implemented an algorithm which generates the full set of model equations based on the number, structure and parameters supplied for the individual LCC and RyR models. As described above, rapid equilibrium for Ca^{2+} flux in the dyadic space is applied to determine the $[Ca^{2+}]_{ds}$ for each possible LCC-RyR open-closed combination as an algebraic function of only the global variables $[Ca^{2+}]_{SR}$, $[Ca^{2+}]_i$, and V. A general CaRU model consisting of N_{LCC} M_{LCC}-state LCCs and $N_{RyR} M_{RyR}$-state RyRs can therefore be employed.

The minimal coupled LCC-RyR model was derived using an LCC model which does not contain a voltage-dependent inactivation mechanism. As described above, the appropriate balance between voltage- and Ca^{2+}-dependent inactivation of the L-type Ca^{2+} current can only be achieved at present in a local control myocyte model (Fig. 8D). Here we develop a coupled LCC-RyR model which retains these properties of LCC inactivation and can be used as a replacement for the computationally expensive stochastic simulation of locally controlled CICR in the canine myocyte model described earlier. The LCC model consists of five states in the same configuration as those of Fig 9B. These represent the channel when it is not voltage-inactivated. In addition, each of the five states has an analogous voltage-inactivated state, where the voltage dependence of inactivation is that used previously in the stochastic local control model [47]. All voltage-inactivated states are closed states. The result is a 10-state LCC model which incorporates separate mechanisms of voltage- and Ca^{2+}-dependent inactivation. The RyR is modeled with the 4-state model used previously in the canine myocyte [47]. It is assumed that the baseline model contains only one LCC and one RyR per CaRU, yielding a 40-state coupled LCC-RyR model.

Figure 12 shows EC coupling gain (ordinate) as a function of membrane potential (mV, abscissa) for the baseline model (short dashed line). The gain

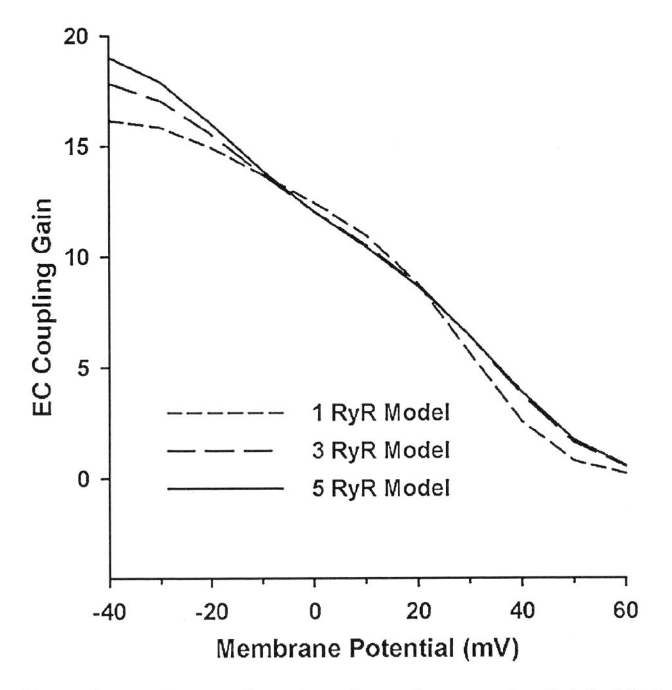

Fig. 12. EC coupling gain as a function of membrane potential (mV) for coupled LCC-RyR models consisting of 1 RyR (*short dashed line*), 3 RyRs (*long dashed line*), and 5 RyRs (*solid line*)

function decreases with increasing potential similar to that shown for the stochastic local control model (Fig. 8C). Since single LCC openings have been shown to activate 4-6 RyRs [51], the inclusion of only one RyR is a model reducing simplification. To test the validity of this model reduction, Fig. 12 also shows EC coupling gain for models which include 3 RyRs per CaRU (long dashed line) and 5 RyRs per CaRU (solid line). The 3-RyR model yields a 200-state coupled LCC-RyR model and the 5-RyR model yields a 560-state coupled LCC-RyR model. Parameters of each model were adjusted such that all models generated nearly identical APs (with duration of ~300 ms at 1 Hz pacing interval) and such that the total open channel Ca^{2+} flux summed over all RyRs in a CaRU was conserved. The results demonstrate that under these conditions there is little variation in EC coupling gain as a function of the number RyRs in the CaRU and therefore justify the choice of one RyR per CaRU for the baseline model. These findings are consistent with experiments that have indicated that RyRs are functionally coupled and may gate synchronously [14].

Figure 13 demonstrates the ability of the baseline coupled LCC-RyR model, when incorporated into the whole cell model of Greenstein and Winslow [47], to reconstruct action potentials and Ca^{2+} transients of normal

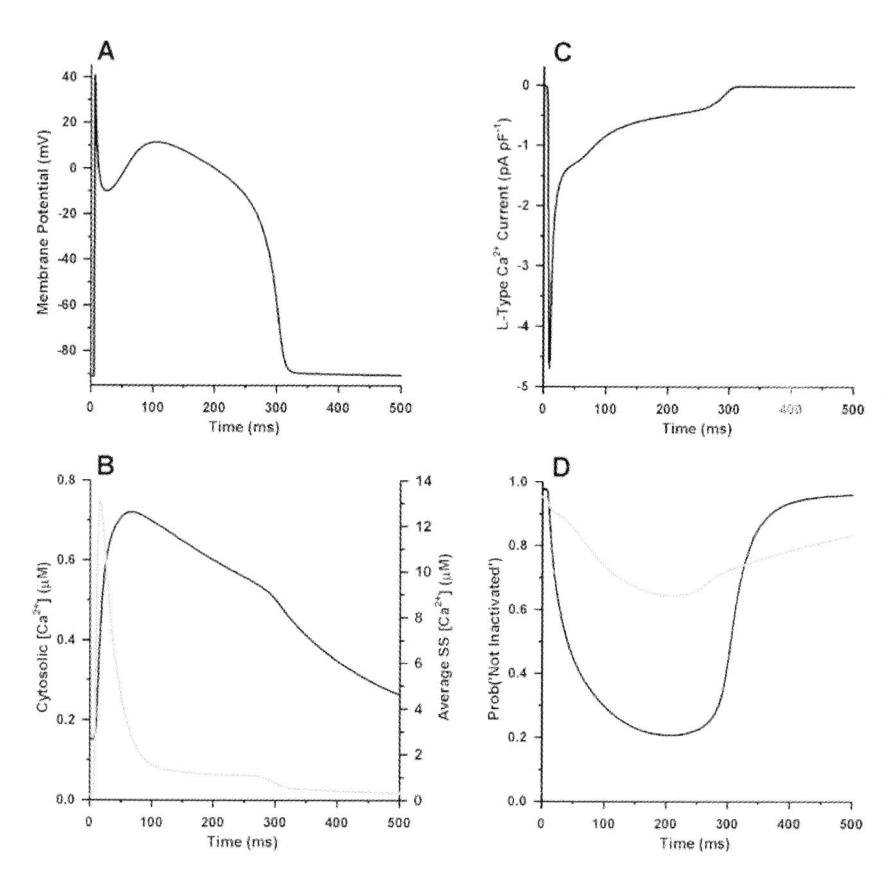

Fig. 13. Whole cell properties of the local control canine myocyte model employing the baseline (1 RyR) coupled LCC-RyR model of CICR. Signals shown are in response to a 1-Hz pulse train, with responses shown in steady-state. (**A**) Action potential simulated under normal conditions. (**B**) Cytosolic (*black line, left axis*) and mean subspace (*gray line, Right axis*) Ca^{2+} concentrations corresponding to the AP simulated in panel A. (**C**) L-type Ca^{2+} current (I_{CaL}) corresponding to the AP simulated in panel A. (**D**) Fraction of LCCs not Ca^{2+}-inactivated (*black line*) and not voltage-inactivated (*gray line*) underlying I_{CaL} (panel C for the AP simulated in panel A

canine midmyocardial ventricular myocytes. In Fig. 13A a normal 1-Hz steady-state AP is shown, and has similar shape and duration (~300 ms) to that of the stochastic model of Fig. 8D [47]. Figure 13B shows cytosolic (black line) and mean subspace (gray line) Ca^{2+} transients. While the cytosolic Ca^{2+} transient peaks at ~0.75 μm, and lasts longer than the duration of the AP, Ca^{2+} in the subspace reaches ~13 μm on average, and equilibrates to near cytosolic levels rapidly within ~100 ms, similar to that observed in the stochastic local control model [47]. L-type current during the AP is shown in Fig. 13C, and peaks at ~4.7 pA pF^{-1} with a sustained component of ~0.5–0.7 pA pF^{-1} which lasts for nearly the entire duration of the AP. Figure 13D shows the relative balance between the fraction of LCCs *not* voltage-inactivated (gray line) and *not* Ca^{2+}-inactivated (black line) during an AP. These fractions agree with the experimental data of shown in Fig. 3A, and the stochastic local control model simulations of Fig. 8D. These results demonstrate that the salient features of the AP, Ca^{2+} cycling, and the balance of voltage- and Ca^{2+}-mediated inactivation of LCCs can be adequately captured in a local control model which employs a relatively low order (40-state), and therefore highly efficient, coupled LCC-RyR model.

4 Modeling Applications

4.1 AP Duration Regulation in Heart Failure

Heart failure (HF), the most common cardiovascular disorder, is characterized by ventricular dilatation, decreased myocardial contractility and cardiac output. Prevalence in the general population is over 4.5 million, and increases with age group to levels as high as 10%. New cases number approximately 400,000 per year. Patient prognosis is poor, with mortality roughly 15% at one year, increasing to 80% at six years subsequent to diagnosis. It is now the leading cause of Sudden Cardiac Death (SCD) in the U.S., accounting for nearly half of all such deaths.

Failing myocytes exhibit altered patterns of expression of several genes/ proteins involved in shaping the cardiac AP. These changes of expression result in prolongation of AP duration (see experimental data in Fig. 14A) and reduction of Ca^{2+} transient amplitude (see experimental data of Fig. 14B). The molecular basis of these changes is now known. Measurements of whole-cell inward rectifier current I_{K1} show that current density at hyperpolarized membrane potentials is reduced in HF by ~50% in human [61], and by ~40% in dog [62]. Measurements of I_{to1} show that in end-stage HF human and canine tachycardia pacing-induced HF indicate current density is reduced by up to 70% in HF [61–63]. No change was observed in either voltage-dependence or kinetics of the I_{to1} current, only a reduction of channel density. Expression of diverse proteins involved in the processes of EC coupling has also been measured in normal and failing myocytes. These proteins include the SR Ca^{2+}-ATPase encoded by the SERCA2α gene and the sodium-calcium (Na^{+}-Ca^{2+}) exchanger protein encoded by the NCX1 gene. Measurements indicate there is an approximate 50% reduction of SERCA2α mRNA, expressed protein level and direct SR Ca^{2+}-ATPase uptake rate in end-stage HF [64]. There is a 55% increase in NCX1 mRNA levels, and an approximate factor of two increase in Na^{+}-Ca^{2+} exchange activity in end-stage HF [64].

In previous work, we investigated the mechanisms by which AP duration is prolonged in ventricular myocytes isolated from failing end-stage hearts

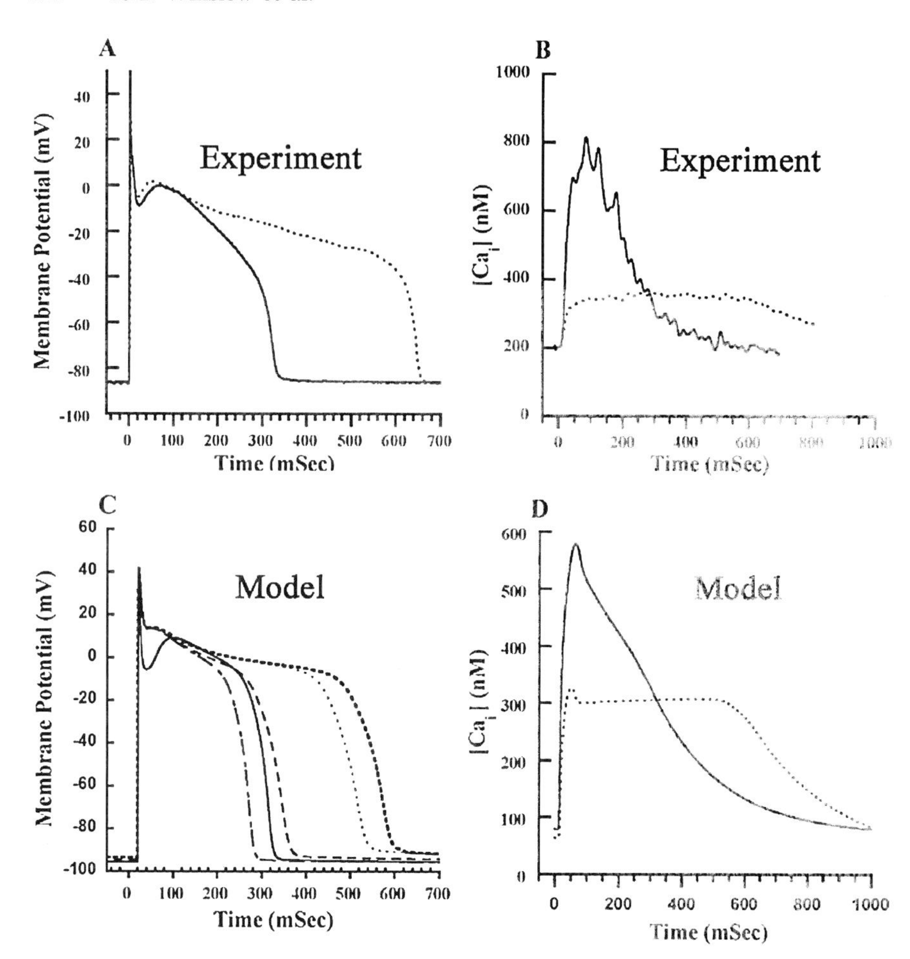

Fig. 14. Model versus experimental action potentials and Ca^{2+} transients. Each action potential and Ca^{2+} transient is in response to a 1 Hz pulse train, with responses measured in the steady-state. (**A**) Experimentally measured membrane potential (mV – ordinate) as a function of time (mSec – abscissa) in normal (*solid*) and failing (*dotted*) canine myocytes. (**B**) Experimentally measured cytosolic Ca^{2+} concentration (nmol/L – ordinate) as a function of time (mSec – abscissa) for normal (*solid*) and failing (*dotted*) canine ventricular myocytes. (**C**) Membrane potential (mV – ordinate) as a function of time (mSec – abscissa) simulated using the normal canine myocyte model (*solid*), and with the successive down-regulation of I_{to1} (*dot-dashed*, 66% down-regulation), I_{K1} (*long-dashed* – down-regulation by 32%), SERCA2α (rightmost *short-dashed* – down-regulation by 62%) and NCX1 (*dotted* – up-regulation by 75%). (**D**) Cytosolic Ca^{2+} concentration (nmol/L – ordinate) as a function of time (mSec – abscissa) simulated using the normal (*solid*) and heart failure (*dotted*) model. Reproduced from Fig. 1 of Winslow et al. [32], with permission

[32,65]. To do so, we formulated a minimal model of altered repolarization and Ca^{2+} handling in ventricular cells from the failing canine heart incorporating: (a) reduced expression of I_{K1} and I_{to1}; (b) down-regulation of the SR Ca^{2+}-ATPase; and (c) up-regulation of the Na^+-Ca^{2+} exchanger. Figures 14C–D demonstrate the ability of the model to reconstruct APs and Ca^{2+} transients measured in both normal and failing canine midmyocardial ventricular myocytes. Figure 14C shows a normal model AP (solid line), and model APs corresponding to the additive effects of sequential down-regulation of I_{to1} (by 62%; dot-dashed line), I_{K1} (by 32%; long-dashed line), and SERCA2α (by 62%; rightmost short-dashed bold line), followed by up-regulation of NCX1 (by 75%; dotted line). Down-regulation of I_{to1} produces a modest *shortening* of AP duration. On first consideration, this seems an anomalous effect, but is one which agrees with recent experiments in canine. The model predicts the mechanism of AP duration shortening is a reduction in driving force due to reduction of the Phase I notch and a reduction in delayed activation of $I_{Ca,L}$ [49]. The additional down-regulation of I_{K1} (long-dashed line) produces modest AP prolongation, consistent with the fact that outward current through I_{K1} is activated primarily at potentials which are hyperpolarized relative to the plateau potential. The most striking result is shown by the short-dashed line in Fig. 14C – significant AP prolongation occurs following down-regulation of SERCA2α. This down-regulation results in a near doubling of AP duration that is similar to that observed experimentally (Fig. 14A).

We have shown that the mechanism of this AP prolongation following down-regulation of SERCA2α involves tight coupling between Ca^{2+} release from the JSR and Ca^{2+}-dependent inactivation of LCCs. Decreases in JSR Ca^{2+} load produced by down-regulation of SERCA2α and up-regulation of NCX1 lead to reduced Ca^{2+} release into the dyadic space. This reduction of the subspace Ca^{2+} transient in turn produces a reduction of calcium-calmodulin (Ca^{2+}/CaM)-dependent inactivation of the L-type Ca^{2+} current. This reduction of inactivation increases the late component of the L-type Ca^{2+} current, prolonging AP duration. Thus, a fundamental hypothesis to emerge from these modeling studies is that tight coupling between JSR Ca^{2+} load, JSR Ca^{2+} release and L-type Ca^{2+} current mediated by Ca^{2+}/CaM-dependent inactivation of LCCs (as shown by the data of Fig. 3) is a critically important modulator of AP duration in HF.

4.2 A General Mechanism for Regulation of AP Duration

The results described above demonstrate that as a consequence of strong Ca^{2+}-dependent inactivation of LCCs by Ca^{2+} release through RyRs, disease processes producing an alteration of JSR Ca^{2+} levels can have a strong effect on AP shape and duration. Recent experimental data has provided more direct evidence that ablation of Ca^{2+}-dependent inactivation of LCCs has a profound effect on the AP. Figure 15A shows APs measured in wild-type (WT) guinea pig ventricular myoyctes versus myocytes in which Ca^{2+}-dependent inactivation of LCCs is ablated by overexpression of a mutant CaM. AP duration in WT cells is approximately 200 ms, whereas that in cells expressing mutant CaM is well over 2500 ms. Figure 15B, obtained using the stochastic local control model, shows that this cellular phenotype is reproduced when Ca^{2+}-dependent inactivation of LCCs is ablated in the model.

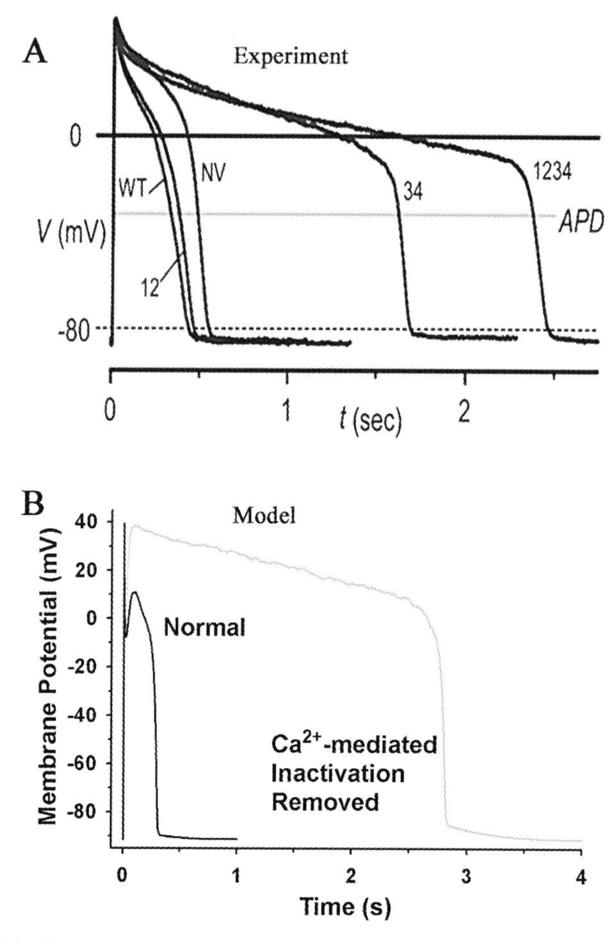

Fig. 15. (**A**) Normal guinea pig ventricular myocyte AP (labeled WT) versus AP measured in myocytes overexpressing muta5CaM (labeled 1234), thus ablating Ca^{2+}-dependent inactivation of LCCs. Reproduced from Fig. 3 of Alseikhan et al. [74], with permission. (**B**) APs simulated using the stochastic local control model with (*black line*) and without (*gray line*) Ca^{2+}-dependent inactivation of LCCs

5 Discussion

Results described in the previous sections indicate that as a result of the research efforts of many laboratories, both experimental investigation as well as mathematical and computational modeling of cardiac EC coupling processes has advanced rapidly. In the following discussion, we address three areas of modeling research which we believe hold particular promise for future development.

5.1 Regulation of Cardiac EC Coupling by Signaling Pathways

Activation of the β-adrenergic (β-AR) signaling pathway enhances cardiac function during stress or exercise through PKA-mediated phosphorylation of target proteins that are directly involved in the process of EC coupling. Targets include LCCs, the SR membrane protein phospholamban (PLB) and RyRs. PKA-mediated phosphorylation of LCCs increases both the fraction of channels available for gating as well as the fraction gating in mode 2 (a gating mode characterized by long-lasting channel openings). The resulting increase in L-type Ca^{2+} current boosts the trigger signal for CICR. Phosphorylation of PLB relieves its inhibitory regulation of the SERCA2α pump, thereby enhancing SR Ca^{2+} uptake, increasing JSR Ca^{2+} content and thus influencing LCC function through Ca^{2+}-dependent inactivation processes (for review, see Bers [66]). As discussed in Sect. 2.2.2, the functional role of PKA-mediated phosphorylation of RyR remains controversial.

Because of its central role in regulation of cardiac contractility, the development of reconstructive and predictive computational models of the β-AR signaling pathway, the actions of this signaling pathway on molecular targets and the consequences of these actions on myocyte function remains an important goal. First steps have been taken. Recently, Greenstein et al. [67] have developed computational models of the effects of PKA-induced phosphorylation of LCCs, PLB and RyRs. The stochastic local control model described in Sect. 3.2 was then used to investigate the ways in which these phosphorylation events regulate properties of EC coupling. Simulations results indicated that characteristic changes in the voltage-dependent EC coupling gain function may be attributed to phosphorylation-induced alterations of specific target proteins. PKA-induced phosphorylation of PLB and the resulting upregulation of SERCA2α activity increased EC coupling gain uniformly as a function of membrane potential by increasing JSR Ca^{2+} levels and thus RyR release flux, similar to the findings of Ginsburg and Bers [68]. Increased LCC open frequency produced a decrease in gain similar to that measured by Song et al. [54]. Increasing RyR Ca^{2+} sensitivity produced an increase in gain similar in shape to that measured by Viatchenko-Karpinski and Györke [69]. These model results suggest that differences in the effects of β-AR stimulation on experimental estimates of the EC coupling gain function reported in the literature may result from differences in the primary downstream targets of β-AR signaling in each of these studies. They also showed that the phosphorylation-induced changes in LCC gating mode (increased gating in Mode 2) produced by either PKA of CaMKII could function to trigger arrhythmias known as early after-depolarizations [67]. Experimental resolution of the actions of β-AR signaling on molecular targets, especially the RyR, and the analysis of the consequences of these actions on the integrative function of the cardiac myoycte remains an important goal for future model development. In addition, Saucerman et al. [70] have recently developed an experimentally-based computational model of the cardiac β-AR signaling pathway. The integration of such signaling models with quantitative descriptions of the effects of phosphorylation of molecular targets, not only in response to PKA but to CaMKII as well, remains an additional future goal.

5.2 Dyadic Ca^{2+} Dynamics

A range of theoretical and computational models predict that Ca^{2+} concentration in the dyad rises to approximately 10–100 μm during the AP. A simple calculation shows that given the estimated dimensions of the dyad (400 nm diameter, 10 nm depth), the number of Ca^{2+} ions yielding these estimated concentrations is very small indeed – a conservative estimate being 10–100 ions per dyad. This fact calls into question core assumptions of each of the three EC coupling models described in this chapter, namely, that Ca^{2+} levels in the dyad involved in triggering CICR may be represented using a single compartment of uniform concentration and that time-varying changes of Ca^{2+} concentration may be described using the laws of mass action. It may be the case that the random motion of Ca^{2+} ions within the space may contribute an important "noise" factor to interactions between RyRs and LCCs. Future modeling efforts must address this issue, making use of tools such as M-Cell (www.mcell.psc.edu) to describe the microenvironment within which the LCCs and RyRs interact as well as the trajectories of individual Ca^{2+} ions within that environment.

5.3 Localized Signaling within Molecular Complexes

Recordings from rat hippocampal neurons have demonstrated the existence of a pre-assembled macromolecular signaling complex which associates the β-AR with the LCC [71]. The complex also contains a G protein, an adenylyl cyclase, protein kinase A, and a phosphatase. It is also known that in cardiac myocytes, a kinase-anchoring protein 15 (AKAP15) co-immunoprecipitates with LCCs [72]. The physical association of the molecular components of β-AR signaling pathway elements suggests that the chain of signaling events from receptor-ligand binding to phosphorylation of the LCC will be determined by interactions between a small number of molecular entities within a highly localized micro-environment. Modeling studies of systems such as these may benefit from an approach which combines molecular dynamic, stochastic and deterministic methods in order to maintain detailed descriptions of local molecular interactions. Furthermore, the need for implementing local molecular interactions poses a unique challenge to the scientific community. More mathematically efficient ways of describing local phenomena are necessary in order to build models that can be used to rapidly explore hypotheses and/or can be incorporated into larger scale multicellular tissue or whole organ models.

Acknowledgements

This work was supported by the NIH (RO1 HL60133, RO1 HL61711, P50 HL52307), the Falk Medical Trust, the Whitaker Foundation, and IBM Corporation.

References

1. Feske, S., Giltnane, J., Dolmetsch, R., Staudt, L.M. and Rao, A., *Gene regulation mediated by calcium signals in T lymphocytes.* Nat Immunol, 2001. **2**: p. 316–24.
2. Lanahan, A. and Worley, P., *Immediate-early genes and synaptic function.* Neurobiol Learn Mem, 1998. **70**(1-2): p. 37–43.
3. Cortassa, S., Aon, M., Marban, E., Winslow, R. and O'Rourke, B., *An integrated model of cardiac mitochondrial energy metabolism and calcium dynamics.* Biophys. J., 2003. **84**: p. 2734–2755.
4. Brette, F. and Orchard, C., *T-tubule function in mammalian cardiac myocytes.* Circ Res, 2003. **92**(11): p. 1182–92.
5. Langer, G.A. and Peskoff, A., *Calcium concentration and movement in the diadic cleft space of the cardiac ventricular cell.* Biophys. J., 1996. **70**: p. 1169–1182.
6. Bers, D.M. and Perez-Reyes, E., *Ca channels in cardiac myocytes: structure and function in Ca influx and intracellular Ca release.* Cardiovasc Res, 1999. **42**: p. 339–60.
7. Zhang, J.F., Ellinor, P.T., Aldrich, R.W. and Tsien, R.W., *Molecular determinants of voltage-dependent inactivation in calcium channels.* Nature, 1994. **372**(6501): p. 97–100.
8. Colecraft, H.M., Alseikhan, B., Takahashi, S.X., Chaudhuri, D., Mittman, S., Yegnasubramanian, V., Alvania, R.S., Johns, D.C., Marban, E. and Yue, D.T., *Novel functional properties of* Ca^{2+} *channel beta subunits revealed by their expression in adult rat heart cells.* J Physiol, 2002. **541**(Pt 2): p. 435–52.
9. 9. Kamp, T.J. and Hell, J.W., *Regulation of cardiac L-type calcium channels by protein kinase A and protein kinase C.* Circ Res, 2000. **87**(12): p. 1095–102.
10. Peterson, B., DeMaria, C., Adelman, J. and Yue, D., *Calmodulin is the* Ca^{2+} *sensor for* Ca^{2+}*-dependent inactivation of L-type calcium channels.* Neuron, 1999. **1999**: p. 549–558.
11. Peterson, B.Z., Lee, J.S., Mulle, J.G., Wang, Y., Leon, M.d. and Yue, D.T., *Critical determinants of* Ca^{2+}*-dependent inactivation within an EF-hand motif of L-type* Ca^{2+} *channels.* Biophys. J., 2000. **78**: p. 1906–1920.

12. Linz, K.W. and Meyer, R., *Control of L-type calcium current during the action potential of guinea-pig ventricular myocytes.* J Physiol (Lond), 1998. **513**(Pt 2): p. 425–42.
13. Lehnart, S.E., Wehrens, X.H., Kushnir, A. and Marks, A.R., *Cardiac ryanodine receptor function and regulation in heart disease.* Ann N Y Acad Sci, 2004. **1015**: p. 144–59.
14. Marx, S.O., Gaburjakova, J., Gaburjakova, M., Henrikson, C., Ondrias, K. and Marks, A., *Coupled gating between cardiac calcium release channels (ryanodine receptors).* Circ. Res., 2001. **88**: p. 1151–1158.
15. Marx, S.O., Reiken, S., Hisamatsu, Y., Jayaraman, T., Burkhoff, D., Rosemblit, N. and Marks, A.R., *PKA phosphorylation dissociates FKBP12.6 from the calcium release channel (ryanodine receptor): defective regulation in failing hearts.* Cell, 2000. **101**: p. 365–76.
16. Xiao, B., Sutherland, C., Walsh, M.P. and Chen, S.R., *Protein kinase A phosphorylation at serine-2808 of the cardiac Ca^{2+}-release channel (ryanodine receptor) does not dissociate 12.6-kDa FK506-binding protein (FKBP12.6).* Circ Res, 2004. **94**: p. 487–95.
17. Wehrens, X.H., Lehnart, S.E., Reiken, S.R. and Marks, A.R., *Ca^{2+}/calmodulin-dependent protein kinase II phosphorylation regulates the cardiac ryanodine receptor.* Circ Res, 2004. **94**(6): p. e61–70.
18. Fabiato, A., *Time and calcium dependence of activation and inactivation of calcium-induced release of calcium from the sarcoplasmic reticulum of a skinned canine cardiac Purkinje cell.* J. Gen. Physiol., 1985. **85**: p. 247–289.
19. Fabiato, A., *Rapid ionic modifications during the aequorin-detected calcium transient in a skinned canine cardiac Purkinje cell.* J Gen Physiol, 1985. **85**: p. 189–246.
20. Fabiato, A., *Simulated calcium current can both cause calcium loading in and trigger calcium release from the sarcoplasmic reticulum of a skinned canine cardiac Purkinje cell.* J Gen Physiol, 1985. **85**: p. 291–320.
21. Wier, W.G., Egan, T.M., Lopez-Lopez, J.R. and Balke, C.W., *Local control of excitation-contraction coupling in rat heart cells.* J. Physiol., **474**: p. 463–471.
22. Bers, D. and Stiffel, V., *Ratio of ryanodine to dihidropyridine receptors in cardiac and skeletal muscle and implications for E-C coupling.* Am. J. Physiol., 1993. **264**(6 Pt 1): p. C1587–C1593.
23. Langer, G.A., *Myocardial Ion Transporters*, in *The Myocardium*, G.A. Langer, Editor. 1997, Academic Press: San Diego. p. 143–179.
24. Blaustein, M.P. and Lederer, W.J., *Sodium/calcium exchange: its physiological implications.* Physiol Rev, 1999. **79**: p. 763–854.
25. Fujioka, Y., Komeda, M. and Matsuoka, S., *Stoichiometry of Na^{+}-Ca^{2+} exchange in inside-out patches excised from guinea-pig ventricular myocytes.* J Physiol, 2000. **523 Pt 2**: p. 339–51.
26. Kang, T.M., Markin, V.S. and Hilgemann, D.W., *Ion fluxes in giant excised cardiac membrane patches detected and quantified with ion-selective microelectrodes.* J Gen Physiol, 2003. **121**: p. 325–47.
27. Grantham, C.J. and Cannell, M.B., *Ca^{2+} influx during the cardiac action potential in guinea pig ventricular myocytes.* Circ. Res., 1996. **79**: p. 194–200.
28. Shannon, T.R., Ginsburg, K.S. and Bers, D.M. *Reverse mode of the SR Ca pump limits SR Ca uptake in permeabilized and voltage clamped myocytes.* in *Cardiac Sarcoplasmic Reticulum Function and Regulation of Contractility.* 1997. Washington, DC: New York Academy of Sciences.

29. Shannon, T.R., Ginsberg, K.S. and Bers, D.M., *SR Ca uptake rate in permeabilized ventricular myocytes is limited by reverse rate of the SR Ca pump.* Biophys. J., 1997. **72**: p. A167.
30. Stern, M., *Theory of excitation-contraction coupling in cardiac muscle.* Biophys J, 1992. **63**: p. 497–517.
31. Jafri, S., Rice, J.J. and Winslow, R.L., *Cardiac Ca^{2+} dynamics: The roles of ryanodine receptor adaptation and sarcoplasmic reticulum load.* Biophys. J., 1998. **74**: p. 1149–1168.
32. Winslow, R.L., Rice, J.J., Jafri, M.S., Marban, E. and O'Rourke, B., *Mechanisms of Altered Excitation-Contraction Coupling in Canine Tachycardia-Induced Heart Failure. II. Model Studies.* Circ. Res., 1999. **84**: p. 571–586.
33. Noble, D., Varghese, A., Kohl, P. and Noble, P., *Improved Guinea-pig ventricular cell model incorporating a diadic space, Ikr and Iks, and length- and tension-dependent processes.* Can. J. Cardiol., 1998. **14**: p. 123–134.
34. Puglisi, J.L., Wang, F. and Bers, D.M., *Modeling the isolated cardiac myocyte.* Prog Biophys Mol Biol, 2004. **85**(2-3): p. 163–78.
35. Faber, G.M. and Rudy, Y., *Action potential and contractility changes in (i) overloaded cardiac myocytes: a simulation study.* Biophys J, 2000. **78**: p. 2392–404.
36. Luo, C.H. and Rudy, Y., *A dynamic model of the cardiac ventricular action potential: I. Simulations of ionic currents and concentration changes.* Circ Res, 1994. **74**: p. 1071–1096.
37. Priebe, L. and Beuckelmann, D.J., *Simulation study of cellular electric properties in heart failure.* Circ Res, 1998. **82**(11): p. 1206–23.
38. Rice, J.J., Jafri, M.S. and Winslow, R.L., *Modeling short-term interval-force relations in cardiac muscle.* Am J Physiol, 2000. **278**: p. H913.
39. Iyer, V., Mazhari, R. and Winslow, R.L., *A computational model of the human left-ventricular epicardial myocyte.* Biophys. J., 2004. **87**: p. in press.
40. Sham, J.S.K., *Ca^{2+} release-induced inactivation of Ca^{2+} current in rat ventricular myocytes: evidence for local Ca^{2+} signalling.* J. Physiol., 1997. **500**: p. 285–295.
41. Bers, D.M., *Excitation Contraction Coupling and Cardiac Contractile Force.* Series in Cardiovascular Medicine. Vol. 122. 1993, Boston: Kluwer Academic Press.
42. Franzini-Armstrong, C., Protasi, F. and Ramesh, V., *Shape, size, and distribution of Ca^{2+} release units and couplons in skeletal and cardiac muscles.* Biophys J, 1999. **77**: p. 1528–39.
43. Rice, J.J., Jafri, M.S. and Winslow, R.L., *Modeling gain and gradedness of Ca^{2+} release in the functional unit of the cardiac diadic space.* Biophys J, 1999. **77**: p. 1871–84.
44. Soeller, C. and Cannell, M.B., *Numerical simulation of local calcium movements during L-type calcium channel gating in the cardiac diad.* Bipphys. J., 1997. **73**: p. 97–111.
45. Cannell, M.B. and Soeller, C., *Numerical analysis of ryanodine receptor activation by L-type channel activity in the cardiac muscle diad.* Biophys. J., 1997. **73**: p. 112–122.
46. Stern, M.D., Song, L.S., Cheng, H., Sham, J.S., Yang, H.T., Boheler, K.R. and Rios, E., *Local control models of cardiac excitation-contraction coupling. A possible role for allosteric interactions between ryanodine receptors.* J Gen Physiol, 1999. **113**: p. 469–89.

47. Greenstein, J.L. and Winslow, R.L., *An integrative model of the cardiac ventricular myocyte incorporating local control of Ca^{2+} release.* Biophys J, 2002. **83**(6): p. 2918–45.
48. Mazhari, R., Greenstein, J.L., Winslow, R.L., Marban, E. and Nuss, H.B., *Molecular interactions between two long-QT syndrome gene products, HERG and KCNE2, rationalized by in vitro and in silico analysis.* Circ Res, 2001. **89**: p. 33–8.
49. Greenstein, J., Po, S., Wu, R., Tomaselli, G. and Winslow, R.L., *Role of the calcium-independent transient outward current Ito1 in action potential morphology and duration.* Circ. Res., 2000. **87**: p. 1026.
50. Keizer, J., Smith, G.D., Ponce-Dawson, S. and Pearson, J.E., *Saltatory propagation of Ca^{2+} waves by Ca^{2+} sparks.* Biophys J, 1998. **75**: p. 595–600.
51. Wang, S.Q., Song, L.S., Lakatta, E.G. and Cheng, H., *Ca^{2+} signalling between single L-type Ca^{2+} channels and ryanodine receptors in heart cells.* Nature, 2001. **410**(6828): p. 592–6.
52. Parker, I., Zang, W.J. and Wier, W.G., *Ca^{2+} sparks involving multiple Ca^{2+} release sites along Z-lines in rat heart cells.* J Physiol (Lond), 1996. **497**(Pt 1): p. 31–8.
53. Song, L.S., Sham, J.S., Stern, M.D., Lakatta, E.G. and Cheng, H., *Direct measurement of SR release flux by tracking 'Ca^{2+} spikes' in rat cardiac myocytes.* J Physiol, 1998. **512 (Pt 3)**: p. 677–91.
54. Song, L.S., Wang, S.Q., Xiao, R.P., Spurgeon, H., Lakatta, E.G. and Cheng, H., *beta-Adrenergic Stimulation Synchronizes Intracellular Ca^{2+} Release During Excitation-Contraction Coupling in Cardiac Myocytes.* Circ Res, 2001. **88**(8): p. 794–801.
55. Matsumoto, M. and Nishimura, T., *Mersenne twister: A 623-dimensionally equidistributed uniform pseudo-random number generator.* ACMTMCS, 1998. **8**: p. 3–30.
56. Hinch, R., Greenstein, J.L., Tanskanen, A.J. and Winslow, R.L., *A simplified local control model of calcium induced calcium release in cardiac ventricular myocytes.* Biophys. J., 2004. **87**: p. 3723–3736.
57. Hinch, R., *A mathematical analysis of the generation and termination of calcium sparks.* Biophys J, 2004. **86**: p. 1293–307.
58. Sobie, E.A., Dilly, K.W., dos Santos Cruz, J., Lederer, W.J. and Jafri, M.S., *Termination of cardiac Ca^{2+} sparks: an investigative mathematical model of calcium-induced calcium release.* Biophys J, 2002. **83**: p. 59–78.
59. Shannon, T.R., Guo, T. and Bers, D.M., *Ca^{2+} scraps: local depletions of free in cardiac sarcoplasmic reticulum during contractions leave substantial Ca^{2+} reserve.* Circ Res, 2003. **93**: p. 40–5.
60. Zahradnikova, A., Kubalova, Z., Pavelkova, J., Gyorke, S. and Zahradnik, I., *Activation of calcium release assessed by calcium release-induced inactivation of calcium current in rat cardiac myocytes.* Am J Physiol Cell Physiol, 2004. **286**: p. C330–41.
61. Beuckelmann, D.J., Nabauer, M. and Erdmann, E., *Alterations of K^+ currents in isolated human ventricular myocytes from patients with terminal heart failure.* Circ. Res., **73**: p. 379–385.
62. Kaab, S., Nuss, H.B., Chiamvimonvat, N., O'Rourke, B., Pak, P.H., Kass, D.A., Marban, E. and Tomaselli, G.F., *Ionic mechanism of action potential prolongation in ventricular myocytes from dogs with pacing-induced heart failure.* Circ Res, 1996. **78**: p. 262–273.

63. Näbauer, M., Beuckelmann, D.J., Überfuhr, P. and Steinbeck, G., *Regional differences in current density and rate-dependent properties of the transient outward current in subepicardial and subendocardial myocytes of human left ventricle.* Circulation, 1996. **93**: p. 168–177.
64. O'Rourke, B., Peng, L.F., Kaab, S., Tunin, R., Tomaselli, G.F., Kass, D.A. and Marban, E., *Mechanisms of altered excitation-contraction coupling in canine tachycardia-induced heart: Experimental studies.* Circ. Res., 1999. **84**: p. 562–570.
65. Winslow, R., Greenstein, J., Tomaselli, G. and O'Rourke, B., *Computational model of the failing myocyte: Relating altered gene expression to cellular function.* Phil. Trans. Roy. Soc. Lond. A, 2001. **359**: p. 1187–1200.
66. Bers, D.M., *Cardiac excitation-contraction coupling.* Nature, 2002. **415**(6868): p. 198–205.
67. Greenstein, J.L., Tanskanen, A.J. and Winslow, R.L., *Modeling the actions of b-adrenergic signaling on excitation-contraction coupling processes.* Ann. N. Y. Acad. Sci., 2004. **1015**: p. 16–27.
68. Ginsburg, K.S., Weber, C.R. and Bers, D.M., *Control of maximum sarcoplasmic reticulum Ca load in intact ferret ventricular myocytes. Effects of thapsigargin and isoproterenol.* J Gen Physiol, 1998. **111**: p. 491–504.
69. Viatchenko-Karpinski, S. and Gyorke, S., *Modulation of the Ca^{2+}-induced Ca^{2+} release cascade by {beta}-adrenergic stimulation in rat ventricular myocytes.* J Physiol (Lond), 2001. **533**: p. 837–848.
70. Saucerman, J.J., Brunton, L.L., Michailova, A.P. and McCulloch, A.D., *Modeling beta-adrenergic control of cardiac myocyte contractility in silico.* J Biol Chem, 2003. **278**(48): p. 47997–8003.
71. Davare, M.A., Avdonin, V., Hall, D.D., Peden, E.M., Burette, A., Weinberg, R.J., Horne, M.C., Hoshi, T. and Hell, J.W., *A beta2 adrenergic receptor signaling complex assembled with the Ca^{2+} channel Cav1.2.* Science, 2001. **293**(5527): p. 98–101.
72. Hulme, J.T., Lin, T.W., Westenbroek, R.E., Scheuer, T. and Catterall, W.A., *Beta-adrenergic regulation requires direct anchoring of PKA to cardiac CaV1.2 channels via a leucine zipper interaction with A kinase-anchoring protein 15.* Proc Natl Acad Sci U S A, 2003. **100**(22): p. 13093–8.
73. Katz, A.M., *Physiology of the Heart.* Second ed. 1992, New York: Raven Press.
74. Alseikhan, B.A., DeMaria, C.D., Colecraft, H.M. and Yue, D.T., *Engineered calmodulins reveal the unexpected eminence of Ca^{2+} channel inactivation in controlling heart excitation.* Proc Natl Acad Sci U S A, 2002. **99**(26): p. 17185–90.

Mathematical Analysis of the Generation of Force and Motion in Contracting Muscle

E. Pate

Department of Mathematics, Washington State University, Pullman, WA 99164
epate@wsu.edu

Key words: myosin, cross-bridge, actin, muscle, model, muscle mechanics

1 Introduction

The forces involved in muscle contraction result from the contractile proteins, myosin and actin. Myosin captures the free energy available from the hydrolysis of adenosine triphosphate (ATP), and via interaction with actin, generates the force and motion necessary for the survival of higher organisms. How this protein-mediated conversion of chemical energy into mechanical energy occurs remains a fundamental, unresolved question in physiology and biophysics. As a problem in thermodynamics, mathematical modeling of this chemomechanical free energy transduction has played an important role in helping to organize the experimental database into a coherent framework. In this chapter, I will discuss basic models that have been used to analyze this really quite remarkable process – the generation of force and motion from a protein-protein interaction involving the ancillary biochemical reaction of nucleotide hydrolysis.

Recent x-ray structures of the protein myosin show that the N-terminus of the protein is composed of a globular motor domain (also termed the catalytic domain) containing the actin- and nucleotide-binding sites. The motor domain is approximately 10 nm in length. An α-helical neck region, approximately 9 nm in length, projects from the motor domain. Skeletal muscle myosin is dimeric with a molecular weight of approximately 520,0000. X-ray, electron microscopic, and secondary structure prediction analyses all suggest the region adjacent to the neck is an α-helical, coiled-coil dimerization domain. The terminal portion of this domain aggregates into the thick filaments seen in muscle ultrastructure. Monomeric G-actin is a globular protein with an approximate diameter of 5–6 nm and a molecular weight of approximately 42,000. Under proper conditions, the G-actin monomers aggregate into a helically arranged, F-actin polymer filament.

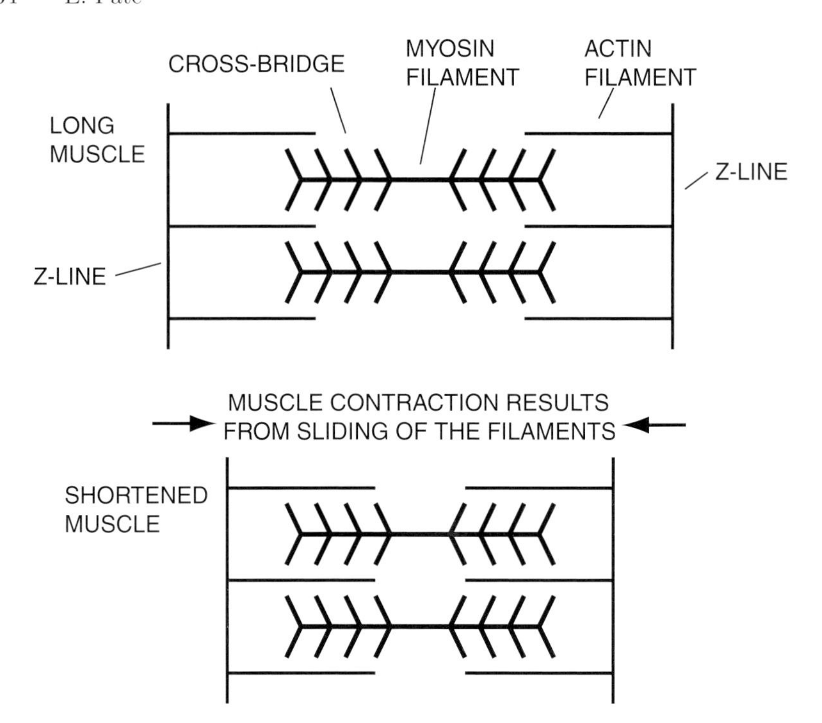

Fig. 1. Cartoon of the relationship between myosin and actin filaments in a sarcomere

The cartoon in Fig. 1 shows the organization of the fundamental contractile unit of muscle, the sarcomere. The sarcomere is an interwoven array of myosin thick filaments, and actin thin filaments. To give relative dimensions, in fast vertebrate skeletal muscle, the myosin filaments are approximately 1.8 μm in length ($1\,\mu\text{m} = 10^{-6}\,\text{m}$). The individual actin filaments are approximately 1 μm long. Myosin cross-bridges project from the thick filaments and can interact with the thin filaments. The most popular hypothesis at this writing is that the functioning portion of the myosin cross-bridge is the myosin motor domain and the neck, discussed above. The strongest evidence in support of this hypothesis is that the motor domain and neck alone are capable of supporting filament sliding [1]. While interacting with actin, the myosin cross-bridges are capable of producing force and muscle shortening. Early microscopy of muscle demonstrated that shortening does not result from a change in length of the myosin and actin filaments (i.e., a "rubber band" model). Instead, the relatively inextensible filaments slide past each other, with muscle shortening the result of a decrease in the Z-line-to-Z-line distance as diagrammed in Fig. 1 [2, 3]. Other geometrical and size parameters are relevant to understanding our subsequent modeling efforts. Only a few interdigitating filaments have been shown in the planar cartoon of Fig. 1. In skeletal muscle, the thin filament types are arranged in hexagonal arrays around the thick filaments.

The cross-bridges project from the thick filaments in a 6-fold symmetry to accommodate the packing. The rest sarcomere length is approximately 2.5 µm. Individual sarcomeres are approximately 1 µm in diameter. They are chained linearly (Z-line to Z-line) in arrays that may be centimeters in length, termed a myofibril. The myofibrils are organized into parallel bundles, termed muscle fibers, which are 50 µm–100 µm in diameter. With proper magnification, individual muscle fibers can be dissected from skinned muscle preparations (the muscle cell wall has been removed mechanically or chemically). These can then be mounted on a mechanical apparatus specifically designed to measure the mechanical properties of contracting muscle. The muscle fiber has been one of the standard experimental preparations in the study of muscle mechanics. Whole muscle preparations, or single intact muscle cells, have likewise been employed. Parameters given have been for skeletal muscle, and there is variation with muscle type. However, these values remain useful order of magnitude estimates. Unless otherwise noted, magnitudes in the remainder of the chapter will be for fast vertebrate skeletal muscle.

A complete discussion of muscle ultrastructure, myosin structure and mechanics, and the actomyosin biochemical interaction is beyond the scope of this chapter. Excellent detailed reviews are in [4–8]. The book by Bagshaw [9] is singled out as an outstanding basic introduction to muscle structure, biochemistry, and mechanics. It is highly recommended as a place for interested researchers to start. The lengthier book by Woledge et al. [10] provides a more advanced discussion of the relationship between biochemistry, mechanics, and energetics. The book by Howard [11] discusses myosin function in the broader context of other families of motor proteins, and motor function in the cytoskeleton.

2 A.F. Huxley's Cross-Bridge Model

The experimental observation is that muscle contraction results from the sliding of otherwise relatively inextensible actin filaments, driven by cross-bridges that extend from the myosin filaments and interact with the adjacent actin filaments. Using this as our starting point, we develop the model for muscle cross-bridge function originally presented by A.F. Huxley [12]. Although this model was first presented over 45 years ago, it still contains the fundamental ideas of the overwhelming majority of models that have followed.

The concentration of myosin cross-bridges in muscle is approximately 240 µM [9]. This means that a single 1 cm long muscle fiber, 50 µm in diameter, will contain in excess of 2×10^{12} cross-bridges. The repeat of the myosin cross-bridges projecting from the thick filaments is different from the repeat of the actin monomers that polymerize to form the actin thin filaments. The large number of cross-bridges in the fiber experimental preparation, and the fact that different cross-bridges will see different distances to the nearest actin-binding site (assumed to be a discrete multiple of the 5.5 nm actin monomer repeat), suggested a continuum model as a first approximation for Huxley [12]. Furthermore, muscle ultrastructure demonstrates that we have an extensive array of interdigitating, parallel filaments. For modeling, this can then be approximated as two infinitely long parallel actin and myosin filaments as shown in the cartoon in Fig. 2. We let x represent the distance between a reference point on a myosin cross-bridge and the nearest actin binding site. We assume a "single-site" assumption where a myosin cross-bridge has a significant probability of interacting with only the nearest actin. With the continuum assumption, the parameter x may be either positive (e.g. x_3 in Fig. 2) or negative (e.g. x_2 in Fig. 2). Shortening with velocity, $v > 0$, will be taken to imply that attached cross-bridges see a decrease in x with time.

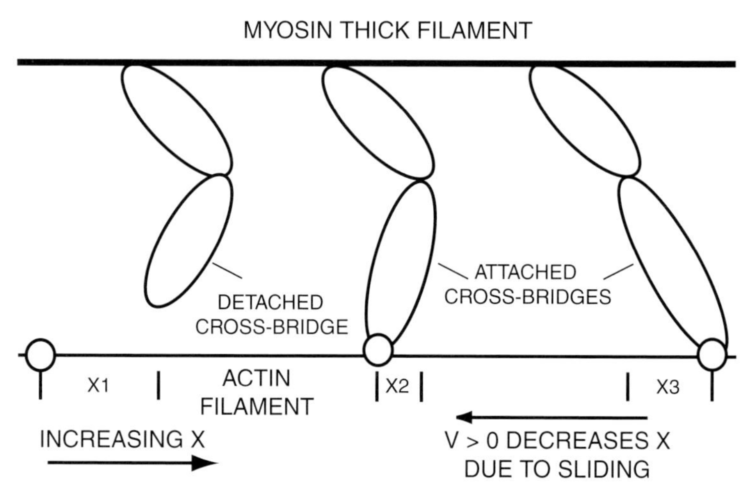

Fig. 2. Cartoon of the Huxley model. For each attached cross-bridge, the reference cross-bridge distortion is idealized as the horizontal distance, x, between the junction of the ellipses representing the cross-bridges and the actin binding site. The value of x increases to the right. Thus $x_2 < 0$ and $x_3 > 0$. Sliding with $v > 0$ decreases the distortion of attached cross-bridges

Let $f(x)$ be the first order (units are s^{-1}), spatially dependent rate of attachment for a myosin cross-bridge that sees the nearest actin binding site at distance x. The quantity, x, is frequently termed the distortion. Let $g(x)$ be the first order detachment rate for a myosin cross-bridge that is attached to an actin binding site at distortion x. Let $n(x,t)$ be the fraction of myosin cross-bridges that see the nearest actin at distortion, x, at time t and are attached. Consider a region of distortions, $[x_0, x_0+\Delta x]$. The attached fraction in this domain can change by four mechanisms. 1. Detached cross-bridges can attach to actin. 2. Attached cross-bridges can detach. 3. Let the filaments slide relative to each other with velocity, $v > 0$. With the sign conventions of Fig. 2, this implies that attached cross-bridges will experience a decrease in distortion with time. Equivalently, at position $x_0 + \Delta x$, there will be a flux of attached cross-bridges into the domain $[x_0,\, x_0 + \Delta x]$. This flux is given by $J(x_0 + \Delta x,\, t) = \rho v n(x_0 + \Delta x,\, t)$, where ρ is the density (per unit length) of cross-bridges on the thick filament. 4. A similar effect occurs at x_0, except that the flux is out of the domain $[x_0 + \Delta x]$, and $J(x_0,\, t) = -\rho v n(x_0,\, t)$. If $n(x,t)$ is the attached fraction, then $1 - n(x,t)$ is the detached fraction. Considering effects 1–4, the balance law giving the time rate of change of the fraction of attached cross-bridges in the domain $[x_0,\, x_0 + \Delta x]$ is

$$\begin{aligned}\frac{\partial}{\partial t}\int_{x_0}^{x_0+\Delta x} \rho n(x,t)dx = &\int_{x_0}^{x_0+\Delta x} f(x)\rho\left[1 - n(x,t)\right]dx \\ &+ \int_{x_0}^{x_0+\Delta x} -g(x)\rho n(x,t)dx \\ &+ J(x_0 + \Delta x) - J(x_0)\end{aligned} \tag{1}$$

i.e. the rate of change = attachment + detachment + flux in − flux out.

Substituting for the definition of J gives a common factor of ρ. Dividing by and applying the mean value theorem for integrals yields

$$\begin{aligned}\frac{\partial n(\xi_1,t)}{\partial t}\Delta x = &f(\xi_2)\left[1 - n(\xi_2,t)\right]\Delta x - g(\xi_3)n(\xi_3,t)\Delta x \\ &+ vn(x_0 + \Delta x, t) - vn(x_0,t)\end{aligned} \tag{2}$$

where $x_0 < \xi_1, \xi_2, \xi_3 < x_0 + \Delta x$. The standard procedure of now dividing by Δx and letting $\Delta x \to 0$, yields a partial derivative in x for the flux, J, terms. Rearranging terms and recognizing that there is nothing special about x_0 so that it can be replaced with x, gives a final balance law of

$$\frac{\partial n(x,t)}{\partial t} - v\frac{\partial n(x,t)}{\partial x} = f(x)\left[1 - n(x,t)\right] - g(x)n(x,t)\ . \tag{3}$$

This hyperbolic partial differential equation, along with initial and boundary conditions, describes the evolution of the attached fraction of cross-bridges in time and space. Note that the left-hand side of the equation is the standard material, or convective derivative, given the somewhat non-standard sign convention adopted for positive velocity.

For comparison with experimental muscle data, the model still requires specification of the rate functions, $f(x)$ and $g(x)$. After considerable trial and error, Huxley settled upon the relationship

$$f(x) = \begin{cases} 0, & x > h \\ \dfrac{f_1 x}{h}, & 0 < x \le h \\ 0, & x \le 0 \end{cases} \tag{4a}$$

and

$$g(x) = \begin{cases} 0, & x > h \\ \dfrac{g_1 x}{h}, & 0 < x \le h \\ g_2, & x \le 0 \end{cases} \tag{4b}$$

Huxley was primarily interested in frog fast vertebrate muscle. Brokaw [13] suggested parameter values of $f_1 = 65\,\mathrm{s}^{-1}$, $g_1 = 15\,\mathrm{s}^{-1}$, $g_2 = 313.5\,\mathrm{s}^{-1}$, and $h = 10\,\mathrm{nm}$ as providing a reasonable fit to the observed contraction velocity at 0°C. Rate functions for these values are shown in Fig. 3, and will be used throughout the text when specific values are employed. Muscle is activated by the release of calcium from the sarcoplasmic reticulum. This would imply time-dependent kinetics as a function of calcium concentration. The Huxley model can viewed as restricting analysis to fully activated conditions.

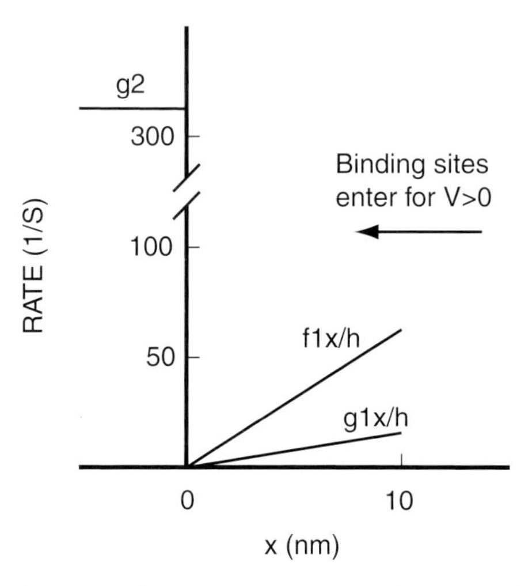

Fig. 3. Kinetic rate functions for cross-bridge attachment and detachment. Values are from Brokaw [13] for frog skeletal muscle

3 Isometric Contraction

Equation (1) is a partial differential equation for cross-bridge function. There are frequently employed experimental protocols that allow the reduction of the model to an ordinary differential equation. The first we will consider is when the muscle, or skinned muscle preparation, is allowed to contract at a fixed length. In this "isometric" mode of contraction, there is no relative sliding of the actin and myosin filaments, $v = 0$, and (1) reduces to the ordinary differential equation in time, with trivial space dependence

$$\frac{dn(x,t)}{dt} = f(x)[1 - n(x,t)] - g(x)n(x,t) \; . \tag{5}$$

Considering the case where all cross-bridges are initially detached, $n(x,0) = 0$, the differential equation for the fraction attached must be solved in the three regions defined by (4a,b), $x \leq 0$, $0 < x \leq h$, and $x > h$. With $f(x) = 0$ outside the region $0 < x \leq h$, we trivially have $n(x,t) = 0$ for $x \leq 0$ and $x > h$.

Combining the solution of

$$\frac{dn(x,t)}{dt} = \frac{f_1(x)}{h}[1 - n(x,t)] - \frac{g_1(x)}{h}n(x,t), \quad n(x,0) = 0, 0 < x \leq h \,, \tag{6}$$

with the zero solution outside this domain, the solution for isometric contraction is given by

$$n(x,t) = \begin{cases} \dfrac{f_1}{f_1 + g_1}\{1 - e^{-(f_1+g_1)xt/h}\}, & 0 < x \leq h \,, \\ 0 & x \leq 0, \; x > h \,. \end{cases} \tag{7}$$

Several observations are relevant. For large time,

$$\lim_{t\to\infty} n(x,t) = \frac{f_1}{f_1 + g_1} \quad \text{for} \quad 0 < x \leq h, \text{ see Fig. 4a.} \tag{8}$$

However, this equilibrium distribution is not reached uniformly. From (7), the time constant for the approach to equilibrium, $(f_1 + g_1)x/h$, is greater for larger x. Cross-bridges with distortions of $x \approx h$ attach significantly faster than cross-bridges with $x \approx 0$. Thus one of the fundamental consequences of the Huxley model is that all cross-bridges are not equal as far as reaction kinetics, and ensuing force production, are concerned.

The function of attached cross-bridges is to produce mechanical force. We have solved for the fraction of attached cross-bridges for a model of two infinitely long parallel filaments. To define the force produced in functioning muscle, the attached fraction must be related to the actual physical system of interdigitating thick and thin filaments. Gordon and co-workers [14] experimentally demonstrated a linear decrease in muscle force with a decrease in thick/thin filament overlap (i.e. an increase in muscle length, see Fig. 1). This implies that the cross-bridges in a sarcomere on individual filaments sum in parallel to produce force. Additionally, we need a mathematical constitutive relationship for the force of an attached cross-bridge. The important experimental observation comes from force changes observed in actively contracting muscle subjected to rapid length changes. The idea is to impose a small length change to the muscle sufficiently rapidly such that significant cross-bridge attachment or detachment does not occur during the length perturbation. Thus one is measuring the mechanics of already attached cross-bridges. If Δt is the time during which the length perturbation is completed, this is equivalent in the Huxley model to requiring $\Delta t \ll (f_1 + g_1)^{-1}$ for stretch, or $\Delta t \ll (f_1 + g_2)^{-1}$ for releases. Experimental data [15] show that for either rapid stretches or rapid releases, the change in mechanical force is linearly proportional to the change in muscle length. This led Huxley to postulate that attached cross-bridges produce force as linear, Hookean springs. Using the notation of Huxley [12], the force produced by an attached cross-bridge with distortion, x, is kx, where k is the elastic force constant. For a muscle with uniform cross-sectional geometry, let A be the cross-sectional area, m be number of cross-bridges per unit volume that can interact with actin, ℓ be the distance between myosin sites and s be the sarcomere length. Then the tension $T(t)$ produced is given by $A(s/2)x$ (average force per cross-bridge), or

$$T(t) = \frac{Asm}{2\ell} \int_{-\ell/2}^{\ell/2} n(x,t)kx dx \; . \tag{9}$$

Myosin filaments are bipolar, with the two ends connected in series. Under conditions of no filament sliding (isometric), each half sarcomere must produce the same force. Thus the requirement of the factor, $s/2$. Due to the fact that muscle preparations are of different diameters, experimental data are generally normalized in terms of force/unit cross-sectional area. Huxley additionally noted that if a myosin cross-bridge has a significant probability of interacting with only the nearest actin binding site (single-site assumption), the integral on $[-\ell/2, \ell/2]$ in (9) can be replaced with a more general integral on $(-\infty, \infty)$, and the tension per unit area, $P(t) = T(t)/A$ is

$$P(t) = \frac{ms}{2\ell} \int_{-\infty}^{\infty} n(x,t)kx dx \; . \tag{10}$$

Letting $n(x,t)$ be defined by the cross-bridge distribution for isometric conditions, (7), substituting into (10), and evaluating the integral, gives

$$\begin{aligned} P(t) = {} & \frac{mksh^2}{4\ell} \left(\frac{f_1}{f_1 + g_1} \right) \left\{ 1 + \frac{2}{(f_1 + g_1)t} e^{-(f_1+g_1)t} \right. \\ & \left. - \frac{2}{[(f_1+g_1)t]^2} \left(1 - e^{-(f_1+g_1)t} \right) \right\} , \end{aligned} \tag{11}$$

and

$$\lim_{t \to \infty} P(t) = \frac{mksh^2}{4\ell} \left(\frac{f_1}{f_1 + g_1} \right) \; . \tag{12}$$

is the final steady-state value of isometric tension, generally represented by P_0.

Protocols involving rapid length changes allow definition of another useful characterization of cross-bridge behavior in actively contracting muscle. Consider the case where a muscle is rapidly stretched by an amount δx per half sarcomere. In other words, each pair of parallel filaments shifts relative to each other by amount δx, again without significant cross-bridge attachment or detachment. Let $S(t)$, termed the mechanical stiffness of the muscle, be the change in force divided by the change in length. Then

$$S(t) = \frac{ms}{2\ell} \left\{ \int_0^h n(x,t)k(x+\delta x)dx - \int_0^h n(x,t)kxdx \right\} / \delta x \ . \tag{13}$$

with

$$\lim_{\delta x \to 0} S(t) = \frac{mks}{2\ell} \int_0^h n(x,t)dx \ . \tag{14}$$

The integral of $n(x,t)$ in space is just the total fraction of attached cross-bridges as a function of time. Thus in the limit of small length perturbations, with all cross-bridges producing force with the same elastic force constant, $S(t)$ is a measure of total cross-bridge attachment.

4 Isotonic Contraction

The first simplification of the balance law for $n(x,t)$, (1), was motivated by the experimental protocol of isometric contraction, with $v = 0$. Another frequently employed experimental protocol is termed isotonic contraction. Here, an actively contracting muscle is allowed to shorten against a constant force, P. The experimental observation is that after a short initial velocity transient, a steady-state shortening at constant velocity is achieved. Additional details of this protocol and representative experimental data can be found in [16]. A.V. Hill [17] first observed that if P_0 is the isometric tension and steady-state shortening velocity is plotted against normalized tension P/P_0, the data were well fit by the hyperbolic relationship

$$V = V_{\text{max}} \frac{(a/P_0)(1 - P/P_0)}{a/P_0 + P/P_0} \ . \tag{15}$$

Here V_{max} is the maximum velocity of shortening and $a/P_0 > 0$ is an additional parameter for the fit. Both V_{max} and a/P_0 vary with muscle type and species. Equation (15) is termed the Hill equation. The physiological range is the portion of the hyperbola that decreases monotonically in $(P/P_0, V)$ space from $(0, V_{\text{max}})$ to $(1, 0)$, see Fig. 5. This relationship implies that muscle shortening works on a "there's no free lunch" basis. In order to shorten more rapidly, muscle has to pay the price of a decrease in the force that it can generate. We next investigate the insights the Huxley model provides into the force-velocity relationship.

For constant velocity shortening with $v > 0$, we can assume a steady-state distribution of cross-bridges in time, and the balance law for cross-bridges, (1), reduces to

$$-v\frac{dn(x)}{dx} = f(x) - [f(x) - g(x)]n(x), \text{ with } \lim_{x \to \infty} n(x) = 0 \tag{16}$$

For $v > 0$, binding sites enter from the right in Fig. 3. Thus it is necessary to solve (16) in the three domains of different rate functions. Using the functional values for $f(x)$ and $g(x)$ in (4), the solution for $x > h$ is trivially $n(x) = 0$. For $0 < x \leq h$, the differential equation becomes

$$-v\frac{dn(x)}{dx} = \frac{f_1 x}{h} - \left(\frac{f_1 x}{h} + \frac{g_1 x}{h}\right) n(x) . \tag{17}$$

For boundary condition, we assume continuity of n at $x = h$. In order to normalize with respect to different length muscles, physiologists usually give shortening velocities in terms of muscle length per unit time. Thus we let $v = (s/2)V$, where s is the sarcomere length. It is also customary to define $\phi = (f_1 + g_1)$ h/s. Then the solution for $n(x)$ in the region $0 < x \leq h$ becomes

$$n(x) = \frac{f_1}{f_1 + g_1}\{1 - e^{[(x^2/h^2-1)\varphi/V]}\} . \tag{18}$$

For $x \leq 0$, the differential equation for n, (17) becomes

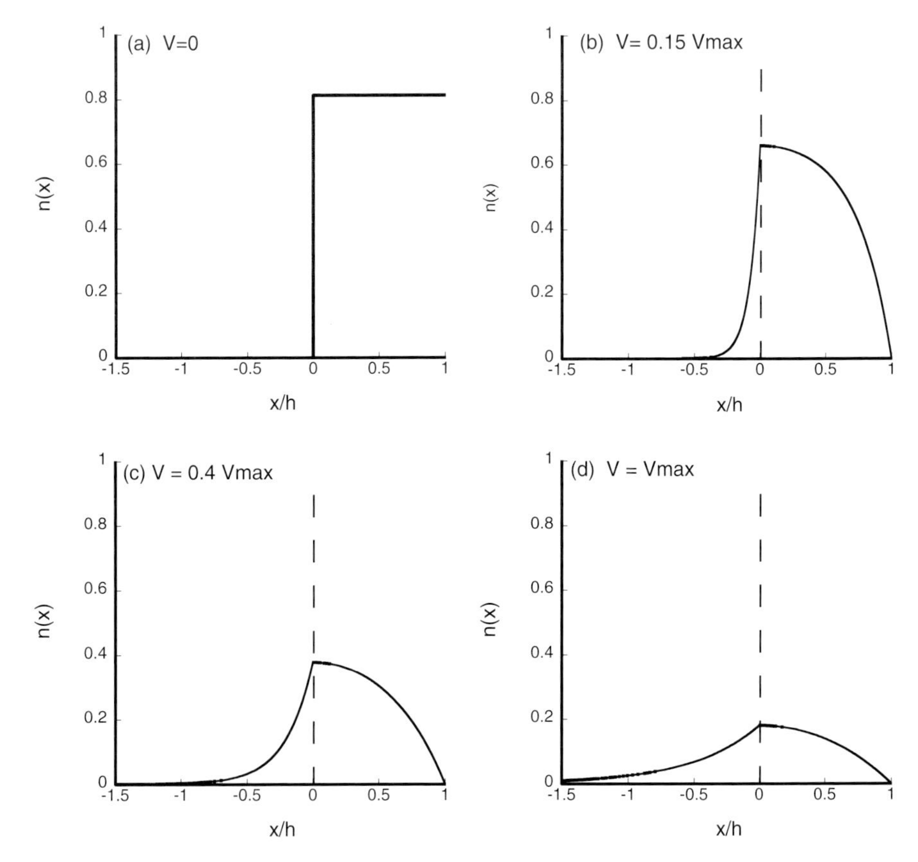

Fig. 4. Cross-bridge distributions, $n(x)$, from (8) (isometric conditions, Panel (**a**) and from (18, 20) for shortening at (**b**) $0.15\,V_{\max}$, (**c**) $0.4\,V_{\max}$, and (**d**) $V_{\max}$

$$-v\frac{dn(x)}{dx} = -g_2 n(x) \ , \tag{19}$$

with continuity at $x = 0$ as the boundary condition. In terms of parameters V and ϕ the solution is

$$n(x) = \frac{f_1}{f_1 + g_1}[1 - e^{(-\varphi/V)}]e^{(2xg_2/sV)} \ . \tag{20}$$

Figure 4 plots $n(x)$ as a function of shortening velocity. In the Huxley model, attached cross-bridges with $x > 0$ produce a positive force supporting active shortening and will be termed powerstroke bridges. Those with $x < 0$, produce a negative, resistive force to active shortening and will be termed dragstroke cross-bridges. For $V = 0$, there is a spatially uniform distribution of powerstroke cross-bridges and no dragstroke bridges (Fig. 4a). Binding sites enter from the right for $V > 0$. Attachment begins when the distance to an actin binding site satisfies $x \leq h$ and the fraction of attached cross-bridges increases with passage through the powerstroke toward $x = 0$. However, the time of passage decreases with increasing velocity and thus the maximum fraction attached in the region $x > 0$ decreases with increasing velocity (Fig. 4b–d). For $x \leq 0$, the flux term in (1) implies that cross-bridges will be carried into the dragstroke region by work done by powerstroke cross-bridges. More rapid cross-bridge detachment begins when $x < 0$, due to the discontinuity in the off-rate function $g(x)$. However, this off-rate is again finite, and with increasing velocity, cross-bridges translate, on average, further into the dragstroke region before detaching. Increasing shortening velocity decreases the net force from positively strained cross-bridges relative to that produced by negatively strained cross-bridges. It had long been recognized that muscles have a maximum shortening velocity, and as formalized in the Hill relation, (15), the net cross-bridge force at $V_{\max}$ is zero. Huxley's key conceptual insight was that for the first time, he provided a mechanical explanation for the existence of $V_{\max}$. $V_{\max}$ is simply that velocity at which the negative force produced by the dragstroke cross-bridges exactly balances the positive force produced by the powerstroke cross-bridges.

However, the Hill relation further suggested a specific hyperbolic, functional relationship at intermediate velocities. Inserting the solution for $n(x)$, (18, 20) into the definition of $P(t)$, (11), evaluating the integral, and observing the value of P_0 from (12) gives the relationship

$$P/P_0 = 1 - \frac{V}{\varphi}\left(1 - e^{-\varphi/V}\right)\left[1 + \frac{1}{2}\left(\frac{f_1}{f_1 + g_1}\right)^2 \frac{V}{\varphi}\right] \ . \tag{21}$$

This transcendental equation relating tension and shortening velocity is the Huxley model equivalent of the hyperbolic Hill equation. Both are plotted in Fig. 5. The figure shows that to within the experimental error, the fits are equivalent. The fundamental difference is that the Hill fit is based upon

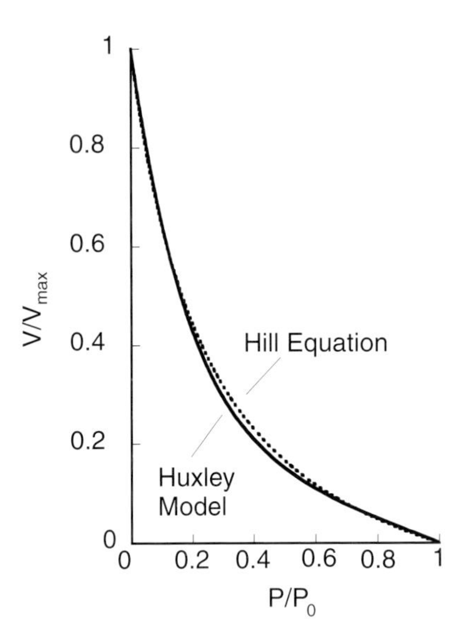

Fig. 5. Force-velocity relationship from the Huxley model (*solid line*) and the Hill equation (*dashed line*). Parameters for the Huxley model are taken from Fig. 3. For the Hill equation, $a/P_0 = 0.25$

heuristic curve fitting of an experimental observation. The Huxley fit follows naturally from the fundamental properties of interacting, linearly elastic cross-bridges.

In his original paper, Huxley also analyzed energy liberation during contraction. Hill [17] suggested that energy liberation (heat + work) increased monotonically with the speed of shortening. Huxley [12] showed that the model also accurately fit the energy liberation data of Hill [17]. However, Aubert [18] later questioned the Hill observation, suggesting instead that energy liberation increased initially with shortening velocity, but for rapid shortening velocities, energy liberation decreased with increasing velocity. More advanced technologies always allow for improved experimental equipment. Hill [19] subsequently agreed that energy liberation decreased at high shortening velocities (greater than $\sim V_{\max}/2$). With current technologies, these remain difficult experiments. Despite its experimental limitations, the energy analysis of Huxley [12] showed for the first time the mathematical framework of how a cross-bridge model can be used to interpret energy liberation. It thus remains extremely informative, and the reader is referred to the original work for details [12]. It should likewise be noted that having developed a model, and then having experimentalists change the experimental data underpinning the model, remains a professional risk for modelers even today.

Since Huxley's original work, a number of modifications have been proposed, while maintaining the same basic conceptual framework. A complete discussion of modeling after 1957 is beyond the scope of this chapter. I limit discussion to the inclusion of improved modeling of the biochemical cycle and of thermodynamic constraints. The original Huxley model considered a single attached and a single detached state. Lymn and Taylor [20] provided the first widely accepted model of the actin-myosin-ATP biochemical interaction, based upon experimental studies of the isolated proteins in solution, which linked cross-bridge biochemical cycling to cross-bridge muscle mechanics. The model contained five states (two detached, three attached) and is shown in Fig. 6. Forward and reverse transitions are allowed between adjacent states, but the working cycle goes in a clockwise direction.

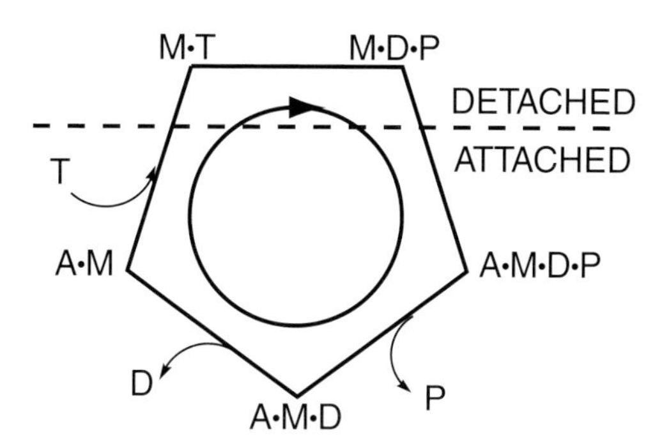

Fig. 6. Kinetic scheme for the actin-myosin-nucleotide interaction. All steps are reversible, but the contractile cycle goes in the clockwise direction. A = actin, M = myosin, T = ATP, D = ADP, and P = orthophosphate. Starting at *upper left*, ATP is hydrolyzed on myosin (in the absence of actin) with a transition to a myosin-hydrolysis products state, M·D·P. The binding to actin yields an A·M·D·P state. P is released first, followed by D, resulting in the A·M, rigor complex. The binding of ATP results in the dissociation of the actomyosin complex into the original M·T state. All states above (*below*) the *dashed line* are detached from (attached to) actin. The actual biochemical cycle is more complicated, and the precise determination of all states involved remains unresolved. This cycle can be viewed as common to more complex models (reviewed in [9] and references therein)

The Huxley model can be viewed as lumping the attached states and detached states into single states. If this is the resolution of the experimental data one wishes to explain, this is a valid simplification. On the other hand, inclusion of the entire five-state model allows for the analysis of mechanical effects of variation of substrate, ATP, and hydrolysis products, ADP and phosphate. There is additional evidence suggesting the mechanical and kinetic properties of the individual states have different dependences on the spatial parameter, x (reviewed in [21, 22] and references therein). More realistic biochemical models are necessary to model these types of effects. However, it must also be recognized that additional, distinct states come at a modeling price. For an N-state model, (1) generalizes to a coupled system of N-1 partial differential equations. I note that Huxley [23] demonstrated that a model containing more than two states could overcome the energy liberation problems of the original two-state model.

Another conceptual advance was the work of Eisenberg and co-workers [21]. In their models, multiple, linearly elastic attached cross-bridge states were assumed with different attached states having different values of distortion for the neutral strain position. In particular, initial attachment was in a weakly attached, low-force, pre-powerstroke state, with subsequent transition into a strongly bound powerstroke attached state, which was mechanically comparable to the attached state in the Huxley model. Considering in greater detail the relationship between thermodynamics and mechanics for linearly elastic cross-bridges, the force produced at distortion, x, for a given state i in a multi-state model was given by $F_i(x) = dG_i(x)/dx$, where $G_i(x)$ is the free energy of the attached state, which is quadratic in x. This is homologous to the relationship in classical physics between a linear spring and its parabolic potential energy as a function of extension, x. Eisenberg and co-workers further observed that if $R_{ij}(x)$ and $R_{ji}(x)$ were the forward and reverse transition rates between states i and j, they could not be thermodynamically independent. They must instead be related by $R_{ij}(x)/R_{ji}(x) = \exp[G_i(x) - G_j(x)]/RT$, where R is the Boltzmann constant and T is absolute temperature. Of the forward rate function, the reverse rate function, and the free energy difference between the two states, the Gibbs relationship implied that only two could be independently specified. The third was then automatically fixed. This condition has been termed "thermodynamic consistency" for cross-bridge models. Additional details are provided in the original references.

5 Transient Simulations

Our discussion of modeling has used two common experimental protocols to reduce the Huxley balance law to ordinary differential equations. The full transient equations have been less frequently employed. They are, for example, required for the modeling of the entire activiation, contraction, and relaxation cycle of cardiac muscle. However, the initial, detailed transient application was for analysis of the beating of cilia and flagella. In this system, the motor protein is dynein. It interacts with the microtubule polymer that forms the tail of a flagella or the cilia hair to cause the rhythmic beating motion. Dynein appears to have many mechanical and biochemical characteristics in common with myosin. Accepting myosin-like behavior, the periodic nature of flagellar and cilliary beating requires solution of the full time- and space-dependent equation. Hines and Blum [25,26] were the first to use finite difference schemes to solve the hyperbolic cross-bridge balance law. These and other numerical schemes can likewise be applied to myosin systems.

Brokaw [13] developed an alternative, probabilistic modeling approach. We discuss his work in greater detail, as it ultimately relates back to the analysis of evolving experimental protocols in actomyosin systems. In this modeling approach, one starts with the knowledge of the states of all the cross-bridges in a finite ensemble at time, t. For the original Huxley model, this would be whether the cross-bridge is attached or detached. Parallel actin and myosin filaments are assumed. One then increments a small increment in time, δt, to time $t + \delta t$. This is done as follows. For each cross-bridge, the distance, x, to the nearest actin-binding site is calculated. If the cross-bridge is attached, the probability of detachment in the time step, δt is determined. For the Huxley model this is

$$p_d(x) = g(x)[1 - \exp(-B(x)\delta t)]/B(x), \quad \text{where } B(x) = f(x) + g(x) . \quad (22)$$

If the cross-bridge is detached, the probability of attachment during a time step, δt is calculated,

$$p_a(x) = f(x)[1 - \exp(-B(x)\delta t)]/B(x) . \quad (23)$$

The probability is compared with a computer-generated pseudorandom number and the cross-bridge status (detached or attached) is updated. All remaining cross-bridges are similarly updated. In reality, computational speed is enhanced by generating a table of probabilities as a function of x and then doing a table lookup at each step. The forces generated by all cross-bridges are then balanced against any imposed external load by shifting the modeled actin filament relative to the modeled myosin filament so that the cross-bridge force and imposed load come into balance. Brokaw [27,28] discusses transition probabilities for models involving more than two states.

The advantage of the probabilistic approach was that it was easy to program, and did not require advanced knowledge of finite difference or other approaches for numerically solving partial differential equations. The disadvantages were that it was a first-order scheme, and that one had to either deal with large ensembles of cross-bridges, or to average multiple simulations with smaller numbers of cross-bridges, in order to eliminate statistical noise. This made the probabilistic scheme computationally more time consuming. However, with massive parallelization on the horizon, and the assumption of independently functioning cross-bridges, this computational speed deficiency may not be the case in the future.

The goal is to use modeling to describe how an individual cross-bridge functions. As noted in Sect. 1, the large numbers of cross-bridges in a muscle fiber justifies the use of a continuum model. The disadvantage of the intact muscle preparation is that the properties of an individual cross-bridge must be deduced from an ensemble average of a very large number of independently functioning cross-bridges that are not synchronized in their chemomechanical cycle. One way to examine better the properties of an individual cross-bridge is to use experimental systems that contain only a few cross-bridges. In this situation, probabilistic models will not be just useful; they will be essential. I describe this application next.

6 Analysis of Systems of Small Numbers of Cross-Bridges

The first experimental protocol that allowed examination of the mechanical properties of small numbers of motor molecules was a reconstituted system of motor proteins termed an in vitro assay [29–32]. In this protocol, as generally employed today, a solution of myosin, or a myosin fragment containing the motor domain, is brought in contact with a nitrocellulose coated glass surface. Individual molecules became fixed to the substrate. The number of motors on the surface can be controlled by the concentration in the solution and the contact time. The solution is then exchanged for one containing polymerized actin filaments and ATP. With proper microscopy, the actin filaments can then be visualized gliding like snakes across a lawn of myosin motors. Gliding velocities are comparable to those observed in whole fiber preparations. Photographs, on-line videos, and additional discussion of in vitro assays can be found on the world-wide web at http://www.mih.unibas.ch/Booklet/Booklet96/Chapter2/Chapter2.html (Andreas Bremer, Daniel Stoffler and Ueli Aebi), http://biochem.stanford.edu/spudich/ (James Spudich), and http://physiology.med.uvm.edu/warshaw/TechspgInVitro.html (David Warshaw). An additional advantage of this protocol is that it can be used with myosins that do not form filaments (e.g., intracellular myosins involved in organelle transport).

We consider an application of probabilistic models to this experimental setup. As noted previously, relative sliding of the actin filaments will be a balance of cross-bridge forces and any imposed load. The average force produced by a single myosin cross-bridge is in the picoNewton range ($1\,\text{pN} = 10^{-12}\,\text{N}$) (reviewed in [5,7,8]). There are two imposed loads that need to be considered in the in vitro assay. One is viscosity; the other is inertia. Approximating an actin filament to be a cylinder of radius 10 nm, the viscous resistance to sliding along the long axis at a typical velocity of 1000 nm/sec will be only about $1\,\text{fN}/\mu\text{M}$ ($1\,\text{fN} = 10^{-15}\,\text{N}$) [33]. The viscous resistance is a factor of 100–1000 less than the force of a *single* cross-bridge, and thus viscous forces will be ignored. For a filament moving through a viscous fluid, the Reynolds number, $\text{Re} = UL/v$, is a dimensionless quantity giving the ratio of inertial to viscous forces. Here U is a velocity scale, L is a length scale and v is the kinematic viscosity of the fluid. Take $U = 1000\,\text{nm/sec}$, $L = 1\,\mu\text{m}$, and $v = 0.01\,\text{cm}^2/\text{sec}$ (water). Then $\text{Re} = 10^{-6}$, and inertial force of the filament is even smaller than the already ignored viscous force. In other words, when we update the filament position for a time step, δt, in the probabilistic model, we are in a situation of unloaded filament sliding. After updating the cross-bridges, assume there are $i = 1, \ldots, n$ attached cross-bridges with distortions x_i. Then our zero net force requirement is determined by a translation of the actin filament by an amount, δx, which satisfies

$$\sum_{i=1}^{n} k(x_i - \delta x) = 0, \text{ with solution } \delta x = \left(\sum_{i=1}^{n} x_i \right) \Big/ n \, . \tag{24}$$

If $n = 0$, no translation is made.

Figure 7 shows representative results from simulations using the Huxley two-state model cross-bridge kinetic scheme, as the number of cross-bridges, N, increases. The graininess of the stepping is evident, although for $N = 50$ it is already dramatically reduced. The other unexpected observation is that the average velocity of sliding increases with cross-bridge number (Fig 7). This has been observed experimentally for myosin [32]. We now consider the factors that determine sliding velocity for a small and a large number of cross-bridges, and show that modeling can demonstrate how the interplay of competing effects offers an explanation for velocity increases with increasing cross-bridge number.

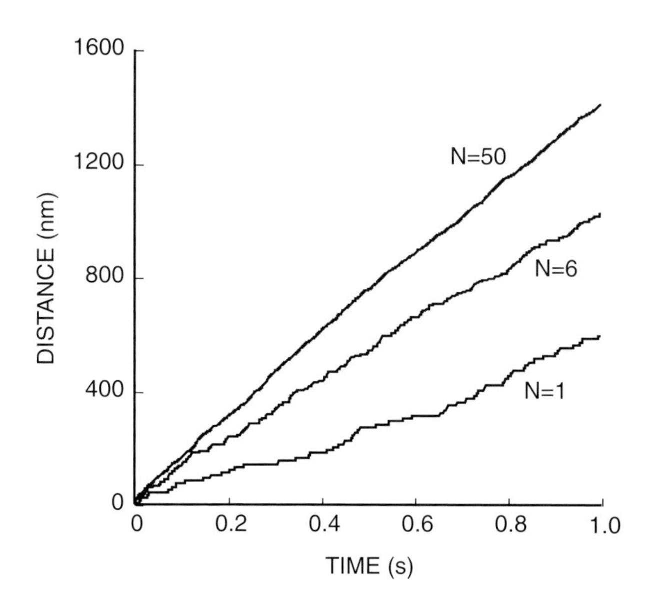

Fig. 7. Sliding distance as a function of time for finite ensembles of Huxley-model cross-bridges. Results for 1, 6, and 50 cross-bridges are shown. Parameters for f_1, g_1, g_2, and h were taken from Fig. 3 (Brokaw [13] model for frog muscle, 0°C)

For the simulations of Fig. 7, myosin cross-bridges were assumed to have the thick filament repeat of 42.9 nm and actin sites a spacing of 10 nm, the powerstroke length. The latter insures retention of single-site, thick-thin filament interaction in even the single cross-bridge, $N = 1$ case, which we consider first. For the parameters used in the simulations, the nearest binding site will be located at $x = h$ after the first step. Using the Huxley kinetic parameters for frog muscle, (4), cross-bridge attachment will occur with rate $f(10\,\mathrm{nm}) = f_1 = 65\,\mathrm{s}^{-1}$. In the presence of negligible resistive forces, translation of the actin filament by a distance of $\delta x = 10\,\mathrm{nm}$, (24) follows. Detachment then occurs with rate $g(0) = g_2 = 313.5\,\mathrm{s}^{-1}$. The mean velocity of sliding will be the mean cycle rate time the powerstroke length. For independent Poisson attachment and detachment processes, this becomes

$$v = (f(h)^{-1} + g(0)^{-1})^{-1} h \,. \tag{25}$$

For our simulations, $f(h) \ll g(0)$, $(f_1 \ll g_2)$ and to a first approximation,

$$v = f(h) h \,. \tag{26}$$

For a large number of cross-bridges, the limit as $N \to \infty$ gives the original Huxley model. Then

$$\begin{aligned} \frac{\partial n(x,t)}{\partial t} - v\frac{\partial n(x,t)}{\partial x} &= -g_2 n(x,t) \,, \quad x \le 0 \\ \frac{\partial n(x,t)}{\partial t} - v\frac{\partial n(x,t)}{\partial x} &= \frac{f_1 x}{h}[1 - n(x,t)] - \frac{g_1 x}{h} n(x,t) \,, \quad 0 < x \le h \end{aligned} \tag{27}$$

with $n(x,t) = 0$ for $x > h$. In order to properly compare the relative sizes of the terms in our analysis, it is necessary to introduce dimensionless, scaled variables. Let $X = x/t$, $T = g_2 t$ be the dimensionless space and time variables. The dimensionless cross-bridge fraction, $n(x,t) \in [0,1]$, and is already scaled. In terms of the new parameters, the balance law for cross-bridges becomes

$$\begin{aligned} \frac{\partial n(X,T)}{\partial T} - V\frac{\partial n(X,T)}{\partial X} &= -n(X,T) \,, \quad X \le 0 \\ \frac{\partial n(X,T)}{\partial T} - V\frac{\partial n(X,T)}{\partial X} &= -RXn(X,T) + HX \,, \quad 0 < X \le 1 \end{aligned} \tag{28}$$

Here the dimensionless velocity $V = v/(g_2 h)$, $R = (f_1 + g_1)/g_2$ is the dimensionless ratio of kinetic constants in the powerstroke and dragstroke force regions, and $H = f_1/g_2$. For steady-state contraction $\partial n/\partial T = 0$, resulting again in ordinary differential equations in X. With $n(1) = 0$ and continuity at $X = 0$, the solution becomes

$$\begin{aligned} n_-(X) &= \frac{H}{R}[1 - e^{-R/2V}]e^{X/V} \,, \quad X \le 0 \\ n_+(X) &= \frac{H}{R}[1 - e^{R(X^2-1)/V}] \,, \quad 0 < X \le 1 \end{aligned} \tag{29}$$

where the $\pm$ subscripts denote the negative and positive distortion region solutions, respectively. The sliding velocity under unloaded conditions will occur when the positive and negative cross-bridge forces balance. With k as the cross-bridge elastic force constant, we require

$$\int_0^1 n_+(X)kXdX + \int_{-\infty}^0 n_-(X)kXdX = 0 \tag{30}$$

Substituting for $n_\pm(X)$ and evaluating the integrals, the dimensionless sliding velocity, V, is the solution of the transcendental equation

$$1 - \frac{2V}{R}(1 - e^{-R/2V})(1 + VR) = 0 \tag{31}$$

The parameter $R = (f_1 + g_1)/g_2 = 0.26$ for the Huxley kinetic parameters. Anticipating that the scaled velocity is O(1), we assume a series expansion for V in terms of powers of R,

$$V = a_0 + a_1 R + a_2 R^2 + \cdots \tag{32}$$

where the a_i are constants. Substitute the power series in R into (31) and expand the exponential term in a Taylor series in $R < 1$. Grouping the terms involving like powers of R, and setting the terms individually to zero, yields a system of algebraic equations that satisfy (31). Solving the equations for the a_i yields

$$V = \frac{1}{2} + \frac{1}{24}R + O(R^2)\,. \tag{33}$$

In terms of dimensioned variables, this becomes (first term only)

$$v = \frac{g_2 h}{2}\,. \tag{34}$$

We can now see the interplay that determines relative sliding velocities at low and high cross-bridge number. For $N = 1$, and the parameters relevant to frog fast skeletal muscle, the rate limiting step for the Huxley for the model is $f(h) = f_1$, the attachment rate at the beginning of the powerstroke. For the case of an infinite number of cross-bridges, the rate-limiting step becomes the detachment rate in the negative force region, $g(x) = g_2$. Due to the magnitudes of f_1 and g_2, the sliding velocity is greater for a large number of cross-bridges.

Note that other parameter ratios can yield different results as a function of cross-bridge number. For example, if $g_2 \ll f_1 + g_1 < f_1$, then for $N = 1$, detachment, $g(0)$, and not attachment, $f(h)$, now dominates in (25). The mean velocity becomes $v = g_2 h$ for the single cross-bridge case. Comparison with the Lymn-Taylor biochemical cycle [20] suggests that this is equivalent to considering a regime in which the concentration of substrate, ATP, is very low [34]. The parameter $R = (f_1 + g_1)/g_2 \gg 1$. The proper expansion for V to solve the transcendental (31) for V is now in powers of $1/R$, and it can be shown that

$$v = \frac{g_2 h}{\sqrt{2}}\,. \tag{35}$$

Thus in this case, the Huxley model predicts that sliding velocity of many cross-bridges will be less than that of a single cross-bridge. Additional details are provided in [34].

In summary, we are now approaching 5 decades after the original work by Huxley. The basic framework of the Huxley model remains the primary foundation for the quantitative interpretation of experimental data. This includes both persons who would claim to be primarily experimentalists and persons who are primarily modelers seeking to add quantitative insight into the experimental database. Although modifications have been made as the experimental database has expanded, the original basic assumptions of elastic attached cross-bridges and spatially dependent cross-bridge kinetics continue to dominate the muscle conversation.

References

1. Toyoshima, Y.Y., S.J. Kron, E.M. McNally, K.R. Niebling, C. Toyoshima, and J.A. Spudich, Myosin subfragment-1 is sufficient to move actin filaments in vitro. Nature, 1987. 328: 536–539.
2. Huxley, A.F. and R. Niedergerke, Structural changes in muscle during contraction; interference microscopy of living muscle fibres. Nature, 1954. 173: 971–973.
3. Huxley, H. and J. Hanson, Changes in the cross-striations of muscle during contraction and stretch and their structural interpretation. Nature, 1954. 173: 973–976.
4. Cooke, R., Force generation in muscle. Curr Opin Cell Biol, 1990. 2: 62–66.
5. Cooke, R., Actomyosin interaction in striated muscle. Physiol Rev, 1997. 77: 671–697.
6. Taylor, E.W., Mechanism of actomyosin ATPase and the problem of muscle contraction. CRC Crit Rev Biochem, 1979. 6: 103–164.
7. Geeves, M.A. and K.C. Holmes, Structural mechanism of muscle contraction. Annu Rev Biochem, 1999. 68: 687–728.
8. Holmes, K.C. and M.A. Geeves, The structural basis of muscle contraction. Philos Trans R Soc Lond B Biol Sci, 2000. 355: 419–431.
9. Bagshaw, C.R., *Muscle Contraction.* Second ed. 1993, London, UK: Chapman and Hall, Ltd. x + 155.
10. Woledge, R., N.A. Curtin, and E. Homsher, *Energetic Aspects of Muscle Contraction.* Monographs of the Physiological Society. Vol. 41. 1985. xiii + 359.
11. Howard, J., *Mechanics of Motor Proteins and the Cytoskeleton.* 2001, Sunderland, MA: Sinauer Associates, Inc. xvi+367.
12. Huxley, A.F., Muscle structure and theories of contraction. Prog Biophys Biophys Chem, 1957. 7: 255–318.
13. Brokaw, C.J., Computer simulation of movement-generating cross-bridges. Biophys J, 1976. 16: 1013–1027.
14. Gordon, A.M., A.F. Huxley, and F.J. Julian, The variation in isometric tension with sarcomere length in vertebrate muscle fibres. J Physiol, 1966. 184: 170–192.
15. Huxley, A.F. and R.M. Simmons, Proposed mechanism of force generation in striated muscle. Nature, 1971. 233: 533–538.
16. Cooke, R., K. Franks, G.B. Luciani, and E. Pate, The inhibition of rabbit skeletal muscle contraction by hydrogen ions and phosphate. J Physiol, 1988. 395: 77–97.
17. Hill, A.V., The heat of shortening and dynamic constraints of muscle. Proc. Royal Soc. London. Series B. Biol. Sci., 1938. 126: 297–318.

18. Aubert, X., *Le couplage energetic de la contraction musculaire*, in *Editions Arsica*. 1956: Brussels, Belgium p. 315.
19. Hill, A.V., The effect of load on the heat of shortening. Proc. Royal Soc. London. Ser. B. Biol. Sci., 1964. 159: 297–318.
20. Lymn, R.W. and E.W. Taylor, Mechanism of adenosine triphosphate hydrolysis by actomyosin. Biochemistry, 1971. 10: 4617–4624.
21. Eisenberg, E. and L.E. Greene, The relation of muscle biochemistry to muscle physiology. Annu Rev Physiol, 1980. 42: 293–309.
22. Pate, E. and R. Cooke, A model of crossbridge action: the effects of ATP, ADP and Pi. J Muscle Res Cell Motil, 1989. 10: 181–196.
23. Huxley, A.F., A note suggesting that the cross-bridge attachment during muscle contraction may take place in two stages. Proc R Soc Lond B Biol Sci, 1973. 183: 83–86.
24. Hill, T.L., Theoretical formalism for the sliding filament model of contraction of striated muscle. Part I. Prog Biophys Mol Biol, 1974. 28: 267–340.
25. Hines, M. and J.J. Blum, Bend propagation in flagella. I. Derivation of equations of motion and their simulation. Biophys J, 1978. 23: 41–57.
26. Hines, M. and J.J. Blum, Bend propagation in flagella. II. Incorporation of dynein cross-bridge kinetics into the equations of motion. Biophys J, 1979. 25: 421–441.
27. Brokaw, C.J. and D. Rintala, Computer simulation of flagellar movement. V. oscillation of cross-bridge models with an ATP-concentration-dependent rate function. J Mechanochem Cell Motil, 1977. 4: 205–232.
28. Brokaw, C.J., Models for oscillation and bend propagation by flagella. Symp Soc Exp Biol, 1982. 35: 313–338.
29. Sheetz, M.P. and J.A. Spudich, Movement of myosin-coated fluorescent beads on actin cables in vitro. Nature, 1983. 303: 31–35.
30. Spudich, J.A., S.J. Kron, and M.P. Sheetz, Movement of myosin-coated beads on oriented filaments reconstituted from purified actin. Nature, 1985. 315: 584–586.
31. Howard, J., A.J. Hudspeth, and R.D. Vale, Movement of microtubules by single kinesin molecules. Nature, 1989. 342: 154–158.
32. Uyeda, T.Q., S.J. Kron, and J.A. Spudich, Myosin step size. Estimation from slow sliding movement of actin over low densities of heavy meromyosin. J Mol Biol, 1990. 214: 699–710.
33. Cox, R.G., The motion of long slender bodies in a viscous fluid. Part 1. General theory. J. Fluid. Mech., 1970. 44: 791–810.
34. Pate, E. and R. Cooke, Simulation of stochastic processes in motile crossbridge systems. J Muscle Res Cell Motil, 1991. 12: 376–393.

Signal Transduction in Vertebrate Olfactory Receptor Cells

J. Reisert

Department of Neuroscience, PCTB 906B, Johns Hopkins School of Medicine, 725 N. Wolfe Street, Baltimore, MD 21205, USA
jreisert@jhmi.edu

Summary. When exposed to odorants, olfactory receptor neurons respond with the generation of action potentials. This conversion of odorous information in the inhaled air into electrical nerve impulses is accomplished by an intracellular enzymatic cascade, which leads to the opening of ion channels and the generation of a receptor current. The resulting depolarisation of the neuron activates voltage-gated ion channels to trigger action potentials, which are conveyed to the olfactory bulb in the brain. This review summarises the information gained over recent years about the details of olfactory signal transduction, including many biophysical parameters helpful for a quantitative description of olfactory signalling.

1 The Odour-Induced Response in Olfactory Receptor Cells

The sensing of odorants begins with the olfactory receptor neurons in the olfactory epithelium which, in rodents, is arranged in a layered structure in the posterior part of the nasal cavity called the turbinates. The epithelium mainly consists of three cell types: supporting cells, basal cells (from which olfactory receptor cells regenerate), and olfactory receptor neurons (ORNs). These bipolar cells have a single axon that carries action potentials directly to the olfactory bulb in the brain and extend a single dendrite to the epithelial border (see Fig. 3). The dendrite terminates in a swelling that is termed the dendritic knob from which ~20 slender cilia protrude into the mucus lining the nasal cavity [1]. The cellular machinery for olfactory transduction from odorant recognition to the generation of the depolarising current, which triggers action potentials in the cell body, is contained entirely within the cilia.

The resting membrane potential of olfactory receptor neurons ranges from −50 to −70 mV [2–5]. When exposed to odour, ORNs respond with a graded depolarisation and associated generation of action potentials. ORNs have a high input resistance and are electronically very compact cells [6–8]. Hence, even quite a small inward, depolarising current can lead to a sufficiently large depolarisation that generates action potentials [8–11]. Yet, olfactory receptor cells are quiet at rest and fire only a few action potentials at a low basal spike rate between 0.05–3 Hz [2, 12–16]. This is an indication that the transduction mechanism itself is quiet and has a low basal activity when not stimulated.

The current, which is generated by an ORN in response to odour stimulation, can be recorded with the whole-cell voltage clamp technique. In this experiment the intracellular potential of the cell is not allowed to vary; it is "clamped" to −55 mV via the recording pipette (Fig. 1A). This current, often referred to as the receptor current, is inward (defined as a negative current) and increases with increasing stimulus concentration [17, 18]. At low odour concentrations the receptor current terminates quickly after the end of stimulation. The peak current magnitude increases steeply with increasing odour concentration and saturating current levels are reached within a 10-fold increase of odour concentration [19]. This large response can outlast the end of stimulation by several seconds. In a different experimental approach the intracellular voltage is allowed to change and the current, which flows across the cell body membrane in response to odour, is collected by sucking the cell body into the tip of a recording pipette (suction pipette technique). In this case, at low concentrations a receptor current with similar time course is observed

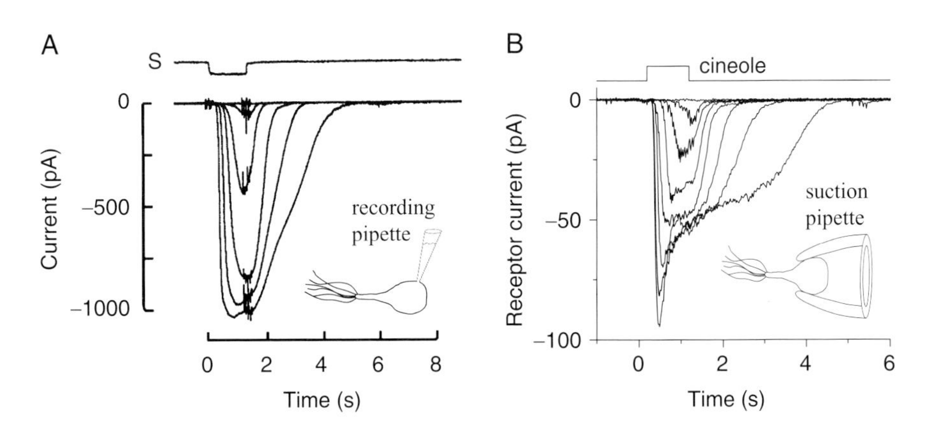

Fig. 1. The odour-induced receptor current to short stimuli. (**A**) The response of an isolated salamander ORN. The odorant amyl acetate was applied at increasing concentrations for 1.2 s as indicated by the stimulus monitor (S) and the response recorded using the whole-cell voltage clamp technique at a holding potential of −55 mV. (**B**) Using a different recording configuration the cell body of a frog ORN was sucked into the tip of a recording pipette and the response to a 1 s exposure to increasing cineole concentrations (also known as eucalyptol) were recorded. Note the characteristic "peak and plateau" kinetics of the current response at higher odour concentrations. Modified from A Firestein et al. [19] and B Reisert and Matthews [11] with permissions from the Journal of Physiology

(Fig. 1B). At high concentrations a characteristic transient current peak of around 0.5 s is recorded followed by a stable plateau current at a lower level, which again can outlast the end of stimulation by several seconds.

With the intracellular voltage free to vary, the cell will generate action potentials driven by the underlying receptor current. In Fig. 1B the current traces were low-pass filtered at 20 Hz to display only the slow receptor current without the fast action potentials.

When exposed to odorants for extended durations (e.g. 30–60 s), ORNs respond with an oscillatory response pattern (Fig. 2). After a larger initial receptor current at the onset of stimulation with associated firing of action potentials, the receptor current returns to a level indistinguishable from zero with in a few seconds. The response adapts to the continuously applied stimulus. Thereafter periodic increases in receptor current (which are quite visible at 100 μM cineole) generate bursts of action potentials with an interburst interval of 5–6 s in the frog and around every second in the mouse [20, 21]. At very high odour concentrations the receptor current does not return to levels near baseline and is present for the entire duration of odour exposure. The receptor current was recorded at a wide bandwidth to display the fast biphasic action potentials.

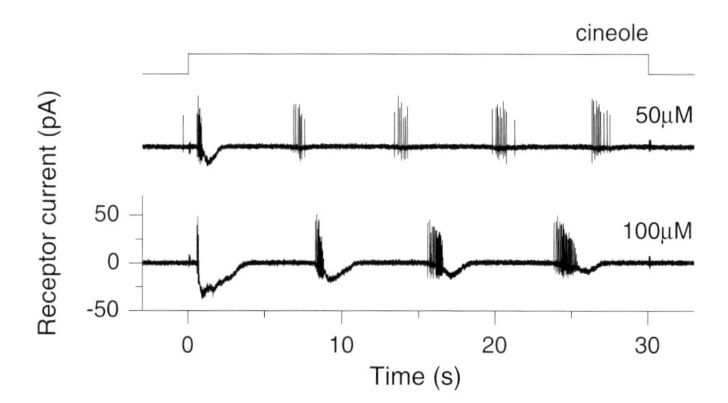

Fig. 2. The oscillating response pattern in response to long stimuli. A frog olfactory receptor neuron was exposed to cineole for 30 s at concentrations indicated next to the traces. The odour-induced electrical response was recorded with the suction pipette technique and filtered at a wide bandwidth to display fast biphasic action potentials and the slower underlying receptor current. Modified from [20] with permission from the Journal of Physiology

2 Olfactory Signal Transduction

2.1 The Olfactory G Protein-Coupled Cascade

Odorants carried into the nose by the inhaled air flow pass the olfactory epithelium and dissolve into the mucus, which lines the nasal cavity. Olfactory signal transduction begins with the binding of an odour molecule to an olfactory receptor protein in the ciliary membrane. These odorant receptors [22] belong to a large gene family of G protein-coupled receptors which contains around 1,000 functional genes in mice [23] and only 350 in humans [24, 25]. The percentage of pseudo genes is much higher in humans compared to rodents. Interestingly, each individual olfactory receptor neuron is thought to express only one receptor type (or at most a few different) [22, 26–28]. This finding introduces a huge puzzle, namely, which of the up to 1,000 odorant receptors recognises which of an almost indefinite number of odorants? Furthermore, it has long been known that an individual receptor neuron, with its one type of odorant receptor, can respond to more than one odorant [19, 29], which indicates that a given odorant receptor can recognise more than one odorant. The process of matching a given odorant to a particular receptor or describing which odorants activate a given odorant receptor is an ongoing process [26, 30–32].

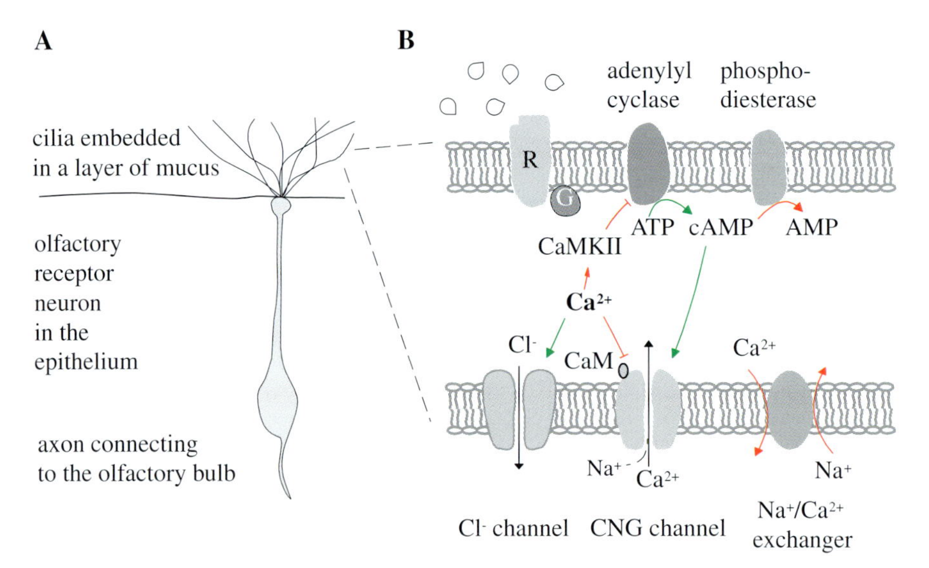

Fig. 3. The olfactory signal transduction cascade. (**A**) The bipolar olfactory receptor neuron with its axon, dendrite and cilia, the site of signal transduction, that protrude into the mucus lining the nasal cavity. (**B**) a schematic cross-section through a cilium with the transduction components. R receptor, G protein, *green* and *red arrows* indicate parts of the pathway involved in the generation or termination of the odour response respectively

Activation of a receptor stimulates the exchange of GDP for GTP on a G protein (Fig. 2) called G_{olf}, which is present on the cytosolic side of the ciliary membrane. Until the bound GTP is hydrolysed to GDP the G protein will stay active and in turn activate adenylyl cyclase type III (AC III). Both, G_{olf} and AC III, are found to be enriched in olfactory cilia [33–36] and their functional significance for olfactory transduction was convincingly demonstrated using mice with a disrupted expression of either protein. Either animal lacks the odorant-induced electroolfactograms (EOGs), which monitor the electrophysiological activity of the olfactory epithelium. Furthermore, both animals nurse poorly or fail olfaction-based behavioural tests [37, 38].

AC III mediated conversion of ATP to cAMP, a cyclic nucleotide, leads to an increase in cAMP in ciliary membrane preparations [39–41]. The level of cAMP production is linearly correlated to the EOG amplitude evoked in response to stimulation by a wide range of different odorants. This indicates that signal transduction mainly occurs with cAMP as the second messenger [42]. Besides cAMP, other second messengers such as IP_3, cGMP and NO have been implicated in olfactory transduction, but little quantitative information is currently available. We would like to point the interested reader towards two comprehensive reviews [43, 44].

2.2 The Olfactory Cyclic Nucleotide-Gated Channel: A Source of Ca^{2+}

The cellular target for the odour-induced rise in intraciliary cAMP was first investigated by Nakamura and Gold. They demonstrated the presence of a cyclic nucleotide-gated (CNG) channel in the ciliary membrane using the excised patch technique [45]. Numerous studies have confirmed that the CNG channel indeed mediates the odour response [46–49]. The channel is rarely found in the membrane of the cell body or dendrite. It is concentrated in the ciliary membrane where it can be found at 100–1,000 times higher density [47, 50]. This substantiates the observation that the cilia are the site of signal transduction. The channel density has been reported to be $70/\mu m^2$ in frog olfactory cilia [51] and 10-fold lower in the rat [52], although much higher values have been reported for both amphibians and rats [50, 53].

The cAMP concentration needed to achieve a half-maximal open probability ($K_{1/2}$) of the CNG channel ranges from 2.5 to 19 µM, depending on the species, and a Hill coefficient of around 2 (see Fig. 5A) has been reported [45, 54–56]. The ciliary concentration of cAMP in the absence of odour stimulation is thought to be around 0.1–0.3 µM, a value just below the threshold of activation [57]. Even at a saturating cAMP concentration the open probability of the channel only reaches 0.7–0.8 [52, 58–60].

In the absence of Ca^{2+} and Mg^{2+} the rat or frog single channel conductance varies between 8–35 pS [50–53, 61]. The CNG channel selects poorly between monovalent cations [46, 54]. However, the presence of physiological concentrations of external divalents introduces a flicker block of the channel and, consequently, a large reduction in single channel conductance [59,62,63]. The block is voltage dependent and most prominent at negative potentials, which is of important functional significance. Without this channel block, opening of a single CNG channel would generate a current sufficient to substantially depolarise a cell with such high input resistance as in the case of olfactory receptor cells. Therefore, even a small fluctuation in the cAMP concentration in the absence of stimulation could trigger action potentials and odour detection would be rendered unreliable [63].

Early indications that Ca^{2+} not only blocks the channel but also permeates [62] were confirmed by Ca^{2+} imaging. When exposed to odour, the ciliary Ca^{2+} concentration rises due to Ca^{2+} influx through the opening CNG channels [64]. The question of how much of the current through the CNG channel is actually carried by monovalents or divalents was investigated by heterologously expressing the olfactory CNG channel in the HEK293 cell line. Simultaneous measurement of the whole cell current and Ca^{2+}-induced fluorescence of the Ca^{2+} sensitive dye fura-2 allowed to estimate the amount of Ca^{2+} which entered during activation of CNG channels and relate it to the overall current. With increasing external Ca^{2+}, the fraction of the total current carried by Ca^{2+} progressively increased and reached 70% at 3 mM Ca^{2+} [65, 66]. Hence, the olfactory CNG channel is a Ca^{2+} channel under physiological conditions and not a non-selective monovalent cation channel.

The olfactory CNG channel is comprised of three subunits. First, a principal subunit called CNGA2, was found in the rat [67], which could be expressed in a heterologous expression system. But, the resulting cAMP gated channel did not resemble the native CNG channel in its biophysical properties. Later, two modulatory channel subunits were found: CNGA4 [68,69] and CNGB1b [60, 70]. Only when all three subunits were expressed together was a channel observed which resembled the native channel [60, 71].

2.3 The Olfactory Ca^{2+}-activated Cl^- Channel: A Large Secondary Current

The finding that the CNG channel is a Ca^{2+} channel, which carries only a small current, raised the question of how the large odour-induced current is generated and which channel or ion carries it. Interestingly, it has been known for a long time that the odour response persists even in the absence of external Na^+, excluding Na^+ as the main excitatory charge carrier [72–75]. Light was shed on this question when cilia were excised from the knobs of frog ORNs using the excised patch technique and exposed to increasing concentrations of Ca^{2+}. A current was generated, which reached half maximal open probability at around 5 μM Ca^{2+} and showed cooperativity of Ca^{2+} binding

(Hill coefficient of 2) [76]. When internal Cl^- was increasingly replaced by gluconate$^-$, the reversal potential shifted to values that closely followed the predictions by the Nernst equation for a Cl^- channel, indicating that Cl^- is the ion, which permeates this channel. Replacing Na^+ by choline$^+$ or Tris$^+$ did not alter the Cl^- current magnitude [76] and Na^+ hardly permeates the channel ($P_{Na}/P_{Cl} = 0.03$ [77]). The single channel conductance of this Ca^{2+}-activated Cl^- channel is small (0.8 pS) [51] and the channel is found at a density of 70 μm^{-2} in frog cilia, a density similar to the CNG channel [51]. Its density in the cell body has not been investigated. The rat olfactory Ca^{2+}-activated Cl^- channel has very similar properties, including the channel density [52].

Is there a role for the Cl^- channel during the odour response and is the resulting current inhibitory or excitatory? When ORNs were stimulated in the presence of the Cl^- channel blocker SITS (see Fig. 4) the odour-induced response decreased, indicating that the Cl^- current is inward (which is equivalent to Cl^- leaving the cilia) and therefore excitatory. In rodents 80% of the odour-induced current [78] and in amphibians 36–65% [79, 80] are carried by Cl^-. Since the Cl^- current is excitatory internal Cl^- must be unusually high to support Cl^- efflux. Indeed, the Cl^- reversal potential was found to be less negative than the resting potential of the cell, indicating a high intracellular Cl^- concentration. Using electrophysiological methods, the Cl^- reversal potential has been shown to be around 0 mV in *Xenopus* [80] and –45 mV in mudpuppy [81]. Alternatively, the internal Cl^- was determined to be between 40 to 80 mM using a variety of methods [82–84]. Furthermore, the Cl^- concentration in the mucus is lower than in interstitial fluid (93 mM in toad mucus, 55 mM in rat mucus [82–85]) thereby aiding Cl^- efflux from the cilia.

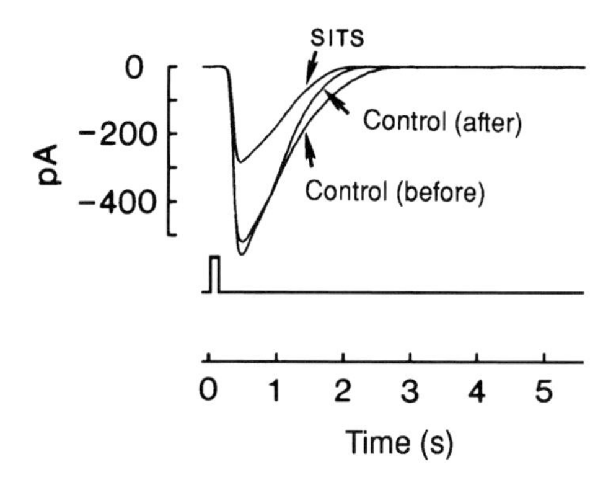

Fig. 4. The contribution of the Cl^- conductance to the odour-induced response. A newt ORN was stimulated as indicated by the solution monitor and the receptor current recorded with the whole-cell voltage clamp technique. Application of the Cl^- channel blocker SITS reversibly reduced the recorded current, indicating that the Cl^- channel conducts a significant portion of the receptor current. From Kurahashi and Yau [79] with permission from Nature

The CNG channel was identified as the sole source of Ca^{2+}, which contributes Ca^{2+} to the opening of Cl^- channels. Application of pharmacological agents that cause Ca^{2+} release from intracellular stores failed to generate ciliary Ca^{2+} transients [86], demonstrating that intraciliary Ca^{2+} stores do not trigger ciliary Ca^{2+} rises. Furthermore, Ca^{2+} signalling in the cell body and the cilia is regulated independently, excluding the cell body as a source of Ca^{2+} [87]. Instead, the ciliary Ca^{2+} has been shown to originate exclusively from the CNG channel, which mainly conducts Ca^{2+} under normal ionic conditions (see above) [64, 87]. It is this small primary Ca^{2+} current which leads to an additional secondary Cl^- current [88].

From a physiological point of view, the Cl^- current has been suggested to serve two roles. First it provides a non-linear, low noise amplification of the CNG current [59, 78]. This is achieved by two means. The single channel conductance of the Cl^- channel is very small, which reduces the noise generated by stochastic opening and closing of individual ion channels. Also, the Ca^{2+} which gates a given Cl^- channel originates not from a single channel, but from several CNG channels, allowing spatial averaging of the Ca^{2+} influx through several CNG channels [52]. With most of the receptor current generated by Cl^- efflux instead of Na^+ influx, the second role is to make the cell's response independent of the mucosal Na^+ concentration, which may vary, especially in amphibians and fish [79, 89]. The Cl^- channel has now been found in many species and always conducts an excitatory current. [21, 52, 76–80, 89, 90].

2.4 Adaptation and Modulation of the Odour Response

An odour-induced receptor current, which is transient and adapts in normal physiological Ringer solution, failed to adapt when Ca^{2+} influx through the CNG channels was minimised by removing external Ca^{2+} [91, 92] or when internal Ca^{2+}-buffering was increased by introducing the Ca^{2+} chelator BAPTA into ORNs [93]. Both demonstrate an important role of Ca^{2+} in the modulation and termination of the odour response.

The CNG channel itself is a target for a Ca^{2+}-mediated negative feedback mechanism. When patches were excised from ORNs and exposed to cAMP in the presence of cytoplasmic Ca^{2+} the CNG current declined, an effect that was attributed to a soluble factor that was later found to be calmodulin (CaM) [91, 94, 95], a small cytoplasmic Ca^{2+}-binding protein present in almost every cell type, including ORNs. Figure 5 shows current-voltage relationships of inside-out membrane patches excised from rat ORNs and currents were activated by applying cAMP at increasing concentrations to the inside of the patch. While 14 μM cAMP yielded a maximal CNG current in the absence of Ca^{2+} and CaM (Fig. 5A), the current to the same cAMP concentration in the presence of 100 μM Ca and 1 μM CaM was less than half maximal (Fig. 5B). The channel's sensitivity to cAMP was reduced and the dose-response relation shifted to higher cAMP concentrations (Fig. 5C) [56, 71]. Therefore, at an intermediate cAMP concentration the CNG channel can open in the absence

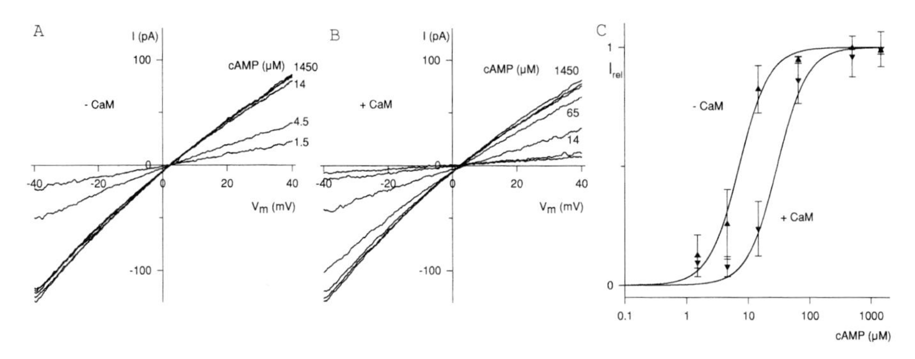

Fig. 5. The effect of Ca-CaM on the CNG channel dose response. Current-voltage relations recorded from membrane patches from rat ORNs to increasing cAMP concentrations in the absence (**A**) and presence (**B**) of 100 µM Ca^{2+} and 1 µM CaM. (**C**) Dose response relations were constructed using the current values at −30 mV and normalised to the maximal current. The data was fitted with the Hill function with $K_{1/2} = 7.1\,\mu$M cAMP and Hill coefficient $n = 1.8$ for "−CaM" and $K_{1/2} = 29.5\,\mu$M, $n = 1.8$ for "+CaM". With permission from Reuter [116]

of Ca-CaM, but will close with a time constant of ~1s upon application of Ca-CaM [71] or, as in the case of an ORN, when intraciliary Ca^{2+} rises due to influx through the CNG channel.

The functional significance of this feedback mechanism to adaptation was demonstrated by Kurahashi and Menini [96]. When ORNs were exposed to two short odour pulses with an increasing recovery time between the two exposures, the response to the second pulse for short recovery times was small but progressively recovered with increasing interpulse intervals. In a second experiment, instead of exposing the cell twice to odour, a light flash was applied that released cAMP from previously loaded caged cAMP. The response to the second light pulse recovered similarly quickly with increasing interflash interval. In the latter case the CNG channel was activated directly without the participation of the upstream transduction cascade limiting the site of adaptation to the CNG channel and downstream components. It was shown using a modelling approach that the negative feedback of Ca-CaM can account for the observed adaptation [96].

For long odour exposures on the time scale of seconds the negative feedback of Ca-CaM on the CNG channel does not suffice to explain experimental results. A receptor current induced by a long exposure to odour declines more quickly than a current of similar peak magnitude generated by the release of cAMP from caged cAMP [97]. This indicates that the generation of cAMP itself is under the influence of a feedback mechanism. Indeed, it has been shown that the activity of AC III is controlled by Ca-CaM dependent kinase II (CaMKII) [98,99], tying the activity of AC III to the intraciliary Ca^{2+} level. Odour stimulation leads to the phosphorylation of AC III at a serine residue, which was prevented in the presence of Ca-CaM kinase inhibitors, as was shown using an antibody specific against phosphorylated AC III [100]. Blocking CaMKII in an intact olfactory receptor cell slows the reduction of the receptor current during prolonged stimulation and reduces adaptation to a subsequent test pulse [93]. Not only the rate of cAMP production, but also its destruction by a phosphodiesterase, seems to be under the control of CaMKII [101–103].

For the odour response to finally terminate not only does the CNG channel have to close, but the Cl^- channel also. Because the Cl^- channel shows only little inactivation during prolonged exposure to Ca^{2+} [52, 76], Ca^{2+} has to be reduced within the cilia to allow the channel to close and the Cl^- current to terminate. The presence of a Na^+/Ca^{2+} exchanger in ORNs has been demonstrated [104, 105] and its functional role was elucidated in ionic substitution experiments. When Na^+ was replaced by Li^+ or choline$^+$, neither of which drive Na^+/Ca^{2+} exchange, a greatly prolonged receptor current was observed that is carried by Cl^- [106]. In Na^+-free solutions, Ca^{2+}, which enters through the CNG channels, cannot be removed by Na^+/Ca^{2+} exchange and the return to pre-stimulus Ca^{2+} levels is slowed. Ca^{2+} remains high and the receptor current cannot terminate, demonstrating the important role of Na^+/Ca^{2+} exchange in response termination. The effect of preventing Na^+/Ca^{2+} exchange is particularly apparent during long odour exposures when the above mentioned oscillatory response pattern can be observed (see Fig. 2). The initial recovery of the receptor current is greatly slowed and the oscillation frequency decreased [20]. A Ca^{2+}-ATPase has also been described in a plasma membrane-rich fraction of olfactory epithelium from salmon [107], indicating that there might be other mechanisms of Ca^{2+} extrusion working in parallel to or at different ciliary Ca^{2+} levels as the Na^+/Ca^{2+} exchanger.

3 Mathematical Approaches in Olfaction

Some aspects of olfaction have been approached with mathematical or modelling tools but the field as a whole is still very much in its infancy. Beginning with the receptor protein itself, pharmacological profiles for individual receptors have been obtained and receptor sequence comparison or investigation of the odour-binding pocket has yielded information about possible odorant receptor interaction sites [108–111]. Alternatively, it was investigated how response properties of individual receptor proteins affect the information coding of the olfactory system as a whole using a model sensory array [112].

The signal transduction cascade and mechanisms leading to the generation of the receptor current were investigated using a quite complex model [113] and action potential firing in response to odour exposure has been studied mathematically by Rospars et al. [114, 115].

References

1. Menco, B.P., *Qualitative and quantitative freeze-fracture studies on olfactory and nasal respiratory structures of frog, ox, rat, and dog. I. A general survey.* Cell Tissue Research, 1980. **207**(2): p. 183–209.
2. Trotier, D. and P. MacLeod, *Intracellular recordings from salamander olfactory receptor cells.* Brain Research, 1983. **268**(2): p. 225–237.
3. Firestein, S., G.M. Shepherd, and F.S. Werblin, *Time course of the membrane current underlying sensory transduction in salamander olfactory receptor neurons.* Journal of Physiology, 1990. **430**(V): p. 135–158.
4. Pun, R.Y.K. and R.C. Gesteland, *Somatic sodium channels of frog olfactory receptor neurons are inactivated at rest.* Pflugers Archiv-European Journal of Physiology, 1991. **418**(5): p. 504–511.
5. Kawai, F., T. Kurahashi, and A. Kaneko, *T-type* Ca^{2+} *channel lowers the threshold of spike generation in the newt olfactory receptor cell.* Journal of General Physiology, 1996. **108**(6): p. 525–535.
6. Trotier, D., *A patch-clamp analysis of membrane currents in salamander olfactory receptor cells.* Pflugers Archiv-European Journal of Physiology, 1986. **407**(6): p. 589–595.
7. Firestein, S. and F. Werblin, *Electrophysiological basis of the response of olfactory receptors to odorant and current stimuli.* Annals of the New York Academy of Sciences, 1987. **510**(V): p. 287–289.
8. Frings, S. and B. Lindemann, *Odorant response of isolated olfactory receptor cells is blocked by amiloride.* Journal of Membrane Biology, 1988. **105**(3): p. 233–243.
9. Maue, R.A. and V.E. Dionne, *Patch-clamp studies of isolated mouse olfactory receptor neurons.* Journal of General Physiology, 1987. **90**(1): p. 95–125.
10. Lynch, J.W. and P.H. Barry, *Action potentials initiated by single channels opening in a small neuron (rat olfactory receptor).* Biophysical Journal, 1989. **55**(4): p. 755–768.
11. Reisert, J. and H.R. Matthews, *Adaptation of the odour-induced response in frog olfactory receptor cells.* Journal of Physiology, 1999. **519**(Sep): p. 801–813.
12. O'Connell, R.J. and M.M. Mozell, *Quantitative stimulation of frog olfactory receptors.* Journal of Neurophysiology, 1969. **32**: p. 51–63.
13. Mathews, D.F., *Response pattern of single neurons in the tortoise olfactory epithelium and olfactory bulb.* Journal of General Physiology, 1972. **60**: p. 166–180.
14. Getchell, T.V., *Unitary responses in frog olfactory epithelium to sterically related molecules at low concentrations.* Journal of General Physiology, 1974. **64**: p. 241–261.
15. Getchell, T.V. and G.M. Shepherd, *Responses of olfactory receptor cells to step pulses of odour at different concentration in the salamander.* Journal of Physiology, 1978. **282**: p. 521–540.
16. Frings, S., S. Benz, and B. Lindemann, *Current recording from sensory cilia of olfactory receptor cells in situ. 2. Role of mucosal* Na^+, K^+, *and* Ca^{2+} *ions.* Journal of General Physiology, 1991. **97**(4): p. 725–747.
17. Firestein, S. and F. Werblin, *Odor-induced membrane currents in vertebrate olfactory receptor neurons.* Science, 1989. **244**(4900): p. 79–82.

18. Kurahashi, T., *Activation by odorants of cation-selective conductance in the olfactory receptor cell isolated from the newt.* Journal of Physiology, 1989. **419**(DEC): p. 177–192.
19. Firestein, S., C. Picco, and A. Menini, *The relation between stimulus and response in olfactory receptor cells of the tiger salamander.* Journal of Physiology, 1993. **468**: p. 1–10.
20. Reisert, J. and H.R. Matthews, *Responses to prolonged odour stimulation in frog olfactory receptor cells.* Journal of Physiology, 2001. **534**(Pt 1): p. 179–191.
21. Reisert, J. and H.R. Matthews, *Response properties of isolated mouse olfactory receptor cells.* Journal of Physiology, 2001. **530**: p. 113–122.
22. Buck, L. and R. Axel, *A novel multigene family may encode odorant receptors: a molecular basis for odor recognition.* Cell, 1991. **65**(1): p. 175–187.
23. Zhang, X. and S. Firestein, *The olfactory receptor gene superfamily of the mouse.* Nature Neuroscience, 2002. **5**(2): p. 124–133.
24. Zozulya, S., F. Echeverri, and T. Nguyen, *The human olfactory receptor repertoire.* Genome Biology, 2001. **2**(6): p. research0018.1–research0018.12.
25. Glusman, G., et al., *The complete human olfactory subgenome.* Genome Research, 2001. **11**(5): p. 685–702.
26. Touhara, K., et al., *Functional identification and reconstitution of an odorant receptor in single olfactory neurons.* Proceedings of the National Academy of Sciences of the United States of America, 1999. **96**(7): p. 4040–4045.
27. Malnic, B., et al., *Combinatorial receptor codes for odors.* Cell, 1999. **96**(5): p. 713–723.
28. Rawson, N.E., et al., *Expression of mRNAs encoding for two different olfactory receptors in a subset of olfactory receptor neurons.* Journal of Neurochemistry, 2000. **75**(1): p. 185–195.
29. Sicard, G. and A. Holley, *Receptor cell responses to odorants: similarities and differences among odorants.* Brain Research, 1984. **292**(2): p. 283–296.
30. Zhao, H.Q., et al., *Functional expression of a mammalian odorant receptor.* Science, 1998. **279**(5348): p. 237–242.
31. Krautwurst, D., K.W. Yau, and R.R. Reed, *Identification of ligands for olfactory receptors by functional expression of a receptor library.* Cell, 1998. **95**(7): p. 917–926.
32. Bozza, T., et al., *Odorant receptor expression defines functional units in the mouse olfactory system.* Journal of Neuroscience, 2002. **22**(8): p. 3033–43.
33. Bakalyar, H.A. and R.R. Reed, *Identification of a specialized adenylyl cyclase that may mediate odorant detection.* Science, 1990. **250**(4986): p. 1403–1406.
34. Jones, D.T. and R.R. Reed, *G_{olf} – an olfactory neuron specific-G protein involved in odorant signal transduction.* Science, 1989. **244**(4906): p. 790–795.
35. Menco, B.P.M., et al., *Ultrastructural localization of olfactory transduction components: the G protein subunit G_{olf}-Alpha and type III adenylyl cyclase.* Neuron, 1992. **8**(3): p. 441–453.
36. Pfeuffer, E., et al., *Olfactory adenylyl cyclase. Identification and purification of a novel enzyme form.* Journal of Biological Chemistry, 1989. **264**(31): p. 18803–18807.
37. Belluscio, L., et al., *Mice deficient in G_{olf} are anosmic.* Neuron, 1998. **20**(1): p. 69–81.

38. Wong, S.T., et al., *Disruption of the type III adenylyl cyclase gene leads to peripheral and behavioral anosmia in transgenic mice.* Neuron, 2000. **27**(3): p. 487–497.
39. Sklar, P.B., R.R.H. Anholt, and S.H. Snyder, *The odorant-sensitive adenylate cyclase of olfactory receptor cells. Differential stimulation by distinct classes of odorants.* Journal of Biological Chemistry, 1986. **261**(33): p. 5538–5543.
40. Pace, U., et al., *Odorant-Sensitive Adenylate Cyclase May Mediate Olfactory Reception.* Nature, 1985. **316**(6025): p. 255–258.
41. Breer, H., I. Boekhoff, and E. Tareilus, *Rapid kinetics of second messenger formation in olfactory transduction.* Nature, 1990. **345**(6270): p. 65–68.
42. Lowe, G., T. Nakamura, and G.H. Gold, *Adenylate cyclase mediates olfactory transduction for a wide variety of odorants.* Proceedings of the National Academy of Sciences of the United States of America, 1989. **86**(14): p. 5641–5645.
43. Schild, D. and D. Restrepo, *Transduction mechanisms in vertebrate olfactory receptor cells.* Physiological Reviews, 1998. **78**(2): p. 429–466.
44. Gold, G.H., *Controversial issues in vertebrate olfactory transduction.* Annual Review of Physiology, 1999. **61**: p. 857–871.
45. Nakamura, T. and G.H. Gold, *A cyclic nucleotide-gated conductance in olfactory receptor cilia.* Nature, 1987. **325**(6103): p. 442–444.
46. Kurahashi, T., *The response induced by intracellular cyclic-AMP in isolated olfactory receptor cells of the newt.* Journal of Physiology, 1990. **430**(V): p. 355–371.
47. Lowe, G. and G.H. Gold, *Contribution of the ciliary cyclic nucleotide-gated conductance to olfactory transduction in the salamander.* Journal of Physiology, 1993. **462**: p. 175–196.
48. Firestein, S., F. Zufall, and G.M. Shepherd, *Single odor-sensitive channels in olfactory receptor neurons are also gated by cyclic nucleotides.* Journal of Neuroscience, 1991. **11**(11): p. 3565–3572.
49. Brunet, L.J., G.H. Gold, and J. Ngai, *General anosmia caused by a targeted disruption of the mouse olfactory cyclic nucleotide-gated cation channel.* Neuron, 1996. **17**(4): p. 681–693.
50. Kurahashi, T. and A. Kaneko, *High density cAMP-gated channels at the ciliary membrane in the olfactory receptor cell.* Neuroreport, 1991. **2**(1): p. 5–8.
51. Larsson, H.P., S.J. Kleene, and H. Lecar, *Noise analysis of ion channels in non-space-clamped cables: Estimates of channel parameters in olfactory cilia.* Biophysical Journal, 1997. **72**(3): p. 1193–1203.
52. Reisert, J., et al., *The Ca-activated Cl channel and its control in rat olfactory receptor neurons.* Journal of General Physiology, 2003. **122**(3): p. 349–364.
53. Kaur, R., et al., *IP_3-gated channels and their occurrence relative to CNG channels in the soma and dendritic knob of rat olfactory receptor neurons.* Journal of Membrane Biology, 2001. **181**(2): p. 91–105.
54. Frings, S., J.W. Lynch, and B. Lindemann, *Properties of cyclic nucleotide-gated channels mediating olfactory transduction. Activation, selectivity, and blockage.* Journal of General Physiology, 1992. **100**(1): p. 45–67.
55. Kurahashi, T. and A. Kaneko, *Gating properties of the cAMP-gated channel in toad olfactory receptor cells.* Journal of Physiology, 1993. **466**: p. 287–302.
56. Chen, T.-Y. and K.-W. Yau, *Direct modulation by Ca^{2+}-calmodulin of cyclic nucleotide-activated channel of rat olfactory receptor neurons.* Nature, 1994. **368**(6471): p. 545–548.

57. Pun, R.Y. and S.J. Kleene, *Contribution of cyclic-nucleotide-gated channels to the resting conductance of olfactory receptor neurons.* Biophysical Journal, 2003. **84**(5): p. 3425–3435.
58. Zufall, F., S. Firestein, and G.M. Shepherd, *Analysis of single cyclic-nucleotide gated channels in olfactory receptor cells.* Journal of Neuroscience, 1991. **11**(11): p. 3573–3580.
59. Kleene, S.J., *High-gain, low-noise amplification in olfactory transduction.* Biophysical Journal, 1997. **73**(2): p. 1110–1117.
60. Bönigk, W., et al., *The native rat olfactory cyclic nucleotide-gated channel is composed of three distinct subunits.* Journal of Neuroscience, 1999. **19**(13): p. 5332–5347.
61. Kolesnikov, S.S., A.B. Zhainazarov, and A.V. Kosolapov, *Cyclic nucleotide-activated channels in the frog olfactory plasma membrane.* FEBS Letters, 1990. **266**: p. 96–98.
62. Zufall, F. and S. Firestein, *Divalent cations block the cyclic nucleotide-gated channel of olfactory receptor neurons.* Journal of Neurophysiology, 1993. **69**(5): p. 1758–1768.
63. Kleene, S.J., *Block by external calcium and magnesium of the cyclic-nucleotide-activated current in olfactory cilia.* Neuroscience, 1995. **66**(4): p. 1001–1008.
64. Leinders-Zufall, T., et al., *Calcium entry through cyclic nucleotide-gated channels in individual cilia of olfactory receptor cells: Spatiotemporal dynamics.* Journal of Neuroscience, 1997. **17**(11): p. 4136–4148.
65. Frings, S., et al., *Profoundly different calcium permeation and blockage determine the specific function of distinct cyclic nucleotide-gated channels.* Neuron, 1995. **15**(1): p. 1690–179.
66. Dzeja, C., et al., *Ca^{2+} permeation in cyclic nucleotide-gated channels.* Embo Journal, 1999. **18**(1): p. 131-144.
67. Dhallan, R.S., et al., *Primary structure and functional expression of a cyclic nucleotide-activated channel from olfactory neurons.* Nature, 1990. **347**(6289): p. 184–187.
68. Bradley, J., et al., *Heteromeric olfactory cyclic nucleotide-gated channels: a subunit that confers increased sensitivity to cAMP.* Proceedings of the National Academy of Sciences of the United States of America, 1994. **91**(19): p. 8890–8894.
69. Liman, E.R. and L.B. Buck, *A second subunit of the olfactory cyclic nucleotide-gated channel confers high sensitivity to cAMP.* Neuron, 1994. **13**(3): p. 611–621.
70. Sautter, A., et al., *An isoform of the rod photoreceptor cyclic nucleotide-gated channel beta subunit expressed in olfactory neurons.* Proceedings of the National Academy of Sciences of the United States of America, 1998. **95**(8): p. 4696–4701.
71. Bradley, J., D. Reuter, and S. Frings, *Facilitation of calmodulin-mediated odor adaptation by cAMP-gated channel subunits.* Science, 2001. **294**(5549): p. 2176–2178.
72. Kleene, S.J. and R.Y.K. Pun, *Persistence of the olfactory receptor current in a wide variety of extracellular environments.* Journal of Neurophysiology, 1996. **75**(4): p. 1386–1391.
73. Yoshii, K. and K. Kurihara, *Role of cations in olfactory reception.* Brain Research, 1983. **274**: p. 239–248.

74. Suzuki, N., *Effects of different ionic invironments on the responses of single olfactory responses in the lamprey.* Comperative Biochemistry and Physiology. A Comperative Physiology, 1978. **61**: p. 461–467.
75. Tucker, D. and T. Shibuya, *A physiologic and pharmacologic study of olfactory receptors.* Cold Spring Harbor symposia on quantitative biology, 1965. **30**: p. 207–15.
76. Kleene, S.J. and R.C. Gesteland, *Calcium-activated chloride conductance in frog olfactory cilia.* Journal of Neuroscience, 1991. **11**(11): p. 3624–3629.
77. Hallani, M., J.W. Lynch, and P.H. Barry, *Characterization of calcium-activated chloride channels in patches excised from the dendritic knob of mammalian olfactory receptor neurons.* Journal of Membrane Biology, 1998. **161**(2): p. 163–171.
78. Lowe, G. and G.H. Gold, *Nonlinear amplification by calcium-dependent chloride channels in olfactory receptor cells.* Nature, 1993. **366**(6452): p. 283–286.
79. Kurahashi, T. and K.-W. Yau, *Co-existence of cationic and chloride components in odorant-induced current of vertebrate olfactory receptor cells.* Nature, 1993. **363**(6424): p. 71–74.
80. Zhainazarov, A.B. and B.W. Ache, *Odor-induced currents in Xenopus olfactory receptor cells measured with perforated-patch recording.* Journal of Neurophysiology, 1995. **74**(1): p. 479–483.
81. Dubin, A.E. and V.E. Dionne, *Action potentials and chemosensitive conductances in the dendrites of olfactory neurons suggest new features for odor transduction.* Journal of General Physiology, 1994. **103**(2): p. 181–201.
82. Reuter, D., et al., *A depolarizing chloride current contributes to chemoelectrical transduction in olfactory sensory neurons in situ.* Journal of Neuroscience, 1998. **18**(17): p. 6623–6630.
83. Nakamura, T., H. Kaneko, and N. Nishida, *Direct measurement of the chloride concentration in newt olfactory receptors with the fluorescent probe.* Neuroscience Letters, 1997. **237**(1): p. 5–8.
84. Kaneko, H., T. Nakamura, and B. Lindemann, *Noninvasive measurement of chloride concentration in rat olfactory receptor cells with use of a fluorescent dye.* American Journal of Physiology, 2001. **280**(6): p. C1387–1393.
85. Chiu, D., T. Nakamura, and G.H. Gold, *Ionic composition of toad olfactory mucus measured with ion-selective microelectrodes.* Chemical Senses, 1988. **13**(4): p. 677–678.
86. Zufall, F., T. Leinders-Zufall, and C.A. Greer, *Amplification of odor-induced Ca^{2+} transients by store-operated Ca^{2+} release and its role in olfactory signal transduction.* Journal of Neurophysiology, 2000. **83**(1): p. 501–512.
87. Leinders-Zufall, T., et al., *Imaging odor-induced calcium transients in single olfactory cilia: Specificity of activation and role in transduction.* Journal of Neuroscience, 1998. **18**(15): p. 5630–5639.
88. Kleene, S.J., *Origin of the chloride current in olfactory transduction.* Neuron, 1993. **11**(1): p. 123–132.
89. Sato, K. and N. Suzuki, *The contribution of a Ca^{2+}-activated Cl^- conductance to amino-acid-induced inward current responses of ciliated olfactory neurons of the rainbow trout.* Journal of Experimental Biology, 2000. **203**(2): p. 253–262.
90. Firestein, S. and G.M. Shepherd, *Interaction of anionic and cationic currents leads to a voltage dependence in the odor response of olfactory receptor neurons.* Journal of Neurophysiology, 1995. **73**(2): p. 562–567.

91. Zufall, F., G.M. Shepherd, and S. Firestein, *Inhibition of the olfactory cyclic nucleotide gated ion channel by intracellular calcium.* Proceedings of the Royal Society of London Series B-Biological Sciences, 1991. **246**(1317): p. 225–230.
92. Kurahashi, T. and T. Shibuya, *Ca^{2+}-dependent adaptive properties in the solitary olfactory receptor cell of the newt.* Brain Research, 1990. **515**(1–2): p. 261–268.
93. Leinders-Zufall, T., M. Ma, and F. Zufall, *Impaired odor adaptation in olfactory receptor neurons after inhibition of Ca^{2+}/calmodulin kinase II.* Journal of Neuroscience, 1999: p. 19:RC19 (1–6).
94. Kramer, R.H. and S.A. Siegelbaum, *Intracellular Ca^{2+} regulates the sensitivity of cyclic nucleotide-gated channels in olfactory receptor neurons.* Neuron, 1992. **9**(5): p. 897–906.
95. Balasubramanian, S., J.W. Lynch, and P.H. Barry, *Calcium-dependent modulation of the agonist affinity of the mammalian olfactory cyclic nucleotide-gated channel by calmodulin and a novel endogenous factor.* Journal of Membrane Biology, 1996. **152**(1): p. 13–23.
96. Kurahashi, T. and A. Menini, *Mechanism of odorant adaptation in the olfactory receptor cell.* Nature, 1997. **385**(6618): p. 725–729.
97. Takeuchi, H. and T. Kurahashi, *Photolysis of caged cyclic AMP in the ciliary cytoplasm of the newt olfactory receptor cell.* Journal of Physiology, 2002. **541**(Pt 3): p. 825–833.
98. Wayman, G.A., S. Impey, and D.R. Storm, *Ca^{2+} inhibition of type III adenylyl cyclase* in vivo. Journal of Biological Chemistry, 1995. **270**(37): p. 21480–21486.
99. Wei, J., G. Wayman, and D.R. Storm, *Phosphorylation and inhibition of type III adenylyl cyclase by calmodulin-dependent protein kinase II* in vivo. Journal of Biological Chemistry, 1996. **271**(39): p. 24231–24235.
100. Wei, J., et al., *Phosphorylation and inhibition of olfactory adenylyl cyclase by CaM kinase II in neurons: a mechanism for attenuation of olfactory signals.* Neuron, 1998. **21**(3): p. 495–504.
101. Yan, C., et al., *Molecular cloning and characterization of a calmodulin-dependent phosphodiesterase enriched in olfactory sensory neurons.* Proceedings of the National Academy of Sciences of the United States of America, 1995. **92**(21): p. 9677–9681.
102. Borisy, F.F., et al., *High-affinity cAMP phosphodiesterase and adenosine localized in sensory organs.* Brain Research, 1993. **610**(2): p. 199–207.
103. Borisy, F.F., et al., *Calcium/calmodulin-activated phosphodiesterase expressed in olfactory receptor neurons.* Journal of Neuroscience, 1992. **12**(3): p. 915–923.
104. Jung, A., et al., *Sodium/calcium exchanger in olfactory receptor neurons of Xenopus laevis.* Neuroreport, 1994. **5**(14): p. 1741–1744.
105. Noe, J., et al., *Sodium/calcium exchanger in rat olfactory neurons.* Neurochemistry International, 1997. **30**(6): p. 523–531.
106. Reisert, J. and H.R. Matthews, *Na^{+}-dependent Ca^{2+} extrusion governs response recovery in frog olfactory receptor cells.* Journal of General Physiology, 1998. **112**(5): p. 529–535.
107. Lo, Y.H., T.M. Bradley, and D.E. Rhoads, *High-affinity Ca^{2+},Mg^{2+}-ATPase in plasma membrane-rich preparations from olfactory epithelium of atlantic salmon.* Biochimica Et Biophysica Acta-Biomembranes, 1994. **1192**(2): p. 153–158.

108. Floriano, W.B., N. Vaidehi, and W.A. Goddard, 3rd, *Making sense of olfaction through predictions of the 3-D structure and function of olfactory receptors.* Chemical Senses, 2004. **29**(4): p. 269–290.
109. Liu, A.H., et al., *Motif-based construction of a functional map for mammalian olfactory receptors.* Genomics, 2003. **81**(5): p. 443–456.
110. Araneda, R.C., et al., *A pharmacological profile of the aldehyde receptor repertoire in rat olfactory epithelium.* Journal of Physiology, 2004. **555**(Pt 3): p. 743–756.
111. Floriano, W.B., et al., *Molecular mechanisms underlying differential odor responses of a mouse olfactory receptor.* Proceedings of the National Academy of Sciences of the United States of America, 2000. **97**(20): p. 10712–10716.
112. Alkasab, T.K., J. White, and J.S. Kauer, *A computational system for simulating and analyzing arrays of biological and artificial chemical sensors.* Chemical Senses, 2002. **27**(3): p. 261–75.
113. Suzuki, N., M. Takahata, and K. Sato, *Oscillatory current responses of olfactory receptor neurons to odorants and computer simulation based on a cyclic AMP transduction model.* Chemical Senses, 2002. **27**(9): p. 789–801.
114. Rospars, J.P., et al., *Spiking frequency versus odorant concentration in olfactory receptor neurons.* Biosystems, 2000. **58**(1–3): p. 133–141.
115. Rospars, J.P., et al., *Relation between stimulus and response in frog olfactory receptor neurons* in vivo. European Journal of Neuroscience, 2003. **18**(5): p. 1135–1154.
116. Reuter, D., *Untersuchungen zur Funktion von* Ca^{2+}*-activierten* Cl^{-} *Kanälen und cAMP-gesteuerten Kationenkanälen bei der Erzeugung des Rezeptorstroms in Riechzellen der Ratte*, in *Institut für Biologische Informationsverarbeitung.* 2000, Universität Köln: Jülich.

Acknowledgements

The author would like to thank A. Roberts and Drs. A. C. Yew, D. P. Dougherty and B. Sandstede for critical and helpful reading of the manuscript and Dr. K.-W. Yau for support.

Mathematical Models of Synaptic Transmission and Short-Term Plasticity

R. Bertram

Department of Mathematics and Kasha Institute of Biophysics,
Florida State University, Tallahassee, Florida 32306

1 Introduction

The synapse is the storehouse of memories, both short-term and long-term, and is the location at which learning takes place. There are trillions of synapses in the brain, and in many ways they are one of the fundamental building blocks of this extraordinary organ. As one might expect for such an important structure, the inner workings of the synapse are quite complex. This complexity, along with the small size of a typical synapse, poses many experimental challenges. It is for this reason that mathematical models and computer simulations of synaptic transmission have been used for more than two decades. Many of these models have focused on the presynaptic terminal, particularly on the role of Ca^{2+} in gating transmitter release (Parnas and Segel, 1981; Simon and Llinás, 1985; Fogelson and Zucker, 1985; Yamada and Zucker, 1992; Aharon et al., 1994; Heidelberger et al., 1994; Bertram et al., 1996; Naraghi and Neher, 1997; Bertram et al., 1999a; Tang et al., 2000; Matveev et al., 2002). The terminal is where neurotransmitters are released, and is the site of several forms of short-term plasticity, such as facilitation, augmentation, and depression (Zucker and Regehr, 2002). Mathematical modeling has been used to investigate the properties of various plasticity mechanisms, and to refine understanding of these mechanisms (Fogelson and Zucker, 1985; Yamada and Zucker, 1992; Bertram et al., 1996; Klingauf and Neher, 1997; Bertram and Sherman, 1998; Tang et al., 2000; Matveev et al., 2002). Importantly, modeling has in several cases been the motivation for new experiments (Zucker and Landò, 1986; Hochner et al., 1989; Kamiya and Zucker, 1994; Winslow et al., 1994; Tang et al., 2000). In this chapter, we describe some of the mathematical models that have been developed for transmitter release and presynaptic plasticity, and discuss how these models have shaped, and have been shaped by, experimental studies.

2 Neurotransmitter Release is Evoked by Ca^{2+} Influx

Neurotransmitters are packaged into small spherical structures called **vesicles**, which have a diameter of 30–50 nm. A presynaptic terminal may contain tens or hundreds of filled vesicles, but typically only a fraction of these are docked to the membrane and ready to be released. The docking stations are called **active zones**, and the docked vesicles form the **readily releasable pool**. When an electrical impulse reaches the synapse, ion channels selective for calcium ions open and Ca^{2+} enters the terminal. Much of this Ca^{2+} is rapidly bound by buffer molecules, but a significant fraction remains free and can bind to Ca^{2+} acceptors associated with a docked vesicle. This causes the vesicle to fuse with the cell membrane, allowing neurotransmitter molecules to diffuse out of the vesicle and into the small space (the **synaptic cleft**) separating the presynaptic and postsynaptic cells. The signal is passed to the postsynaptic cell when the transmitter molecules bind to receptors in the postsynaptic membrane, inducing a change in its membrane potential.

The key role that Ca^{2+} plays in neurotransmitter release raises two questions. First, what is the nature of the Ca^{2+} acceptors that, when bound, result in vesicle fusion? Second, what is the nature of the Ca^{2+} signal? Much effort has gone into answering these questions. It was shown by Dodge and Rahamimoff (1967) that there is a fourth-power relation between the external Ca^{2+} concentration and transmitter release in a neuromuscular junction, suggesting that each release site contains four Ca^{2+} acceptors. Although the identity of the acceptors is still under investigation, there is much evidence suggesting that the protein synaptotagmin plays such a role (Geppert et al., 1994; Goda, 1997). Once bound, does Ca^{2+} unbind from an acceptor rapidly or slowly once the membrane is repolarized and the Ca^{2+} channels close? Some data suggest the latter, providing a mechanism for short-term synaptic enhancement (Stanley, 1986; Bertram et al., 1996).

The central question regarding the nature of the Ca^{2+} signal is whether vesicle fusion is gated by the Ca^{2+} microdomain formed at the mouth of a single open Ca^{2+} channel, by overlapping microdomains from several open channels, or by Ca^{2+} from a large aggregate of channels. This should depend on the location of the Ca^{2+} acceptors. Are they located close to Ca^{2+} channels so that they respond primarily to the plume of Ca^{2+} at the mouth of an open channel, or are some of the acceptors farther away so that they respond to a spatially averaged Ca^{2+} signal? Questions regarding the influence of Ca^{2+} channel and Ca^{2+} acceptor geometry have been studied largely through mathematical modeling, constrained by experimental studies of the timing of transmitter release and the effects of exogenous Ca^{2+} buffers.

3 Primed Vesicles are Located Close to Ca^{2+} Channels

Once transmitter-filled vesicles have been transported to the terminal membrane, they then dock and are primed for release. It is only the primed vesicles that can fuse following a presynaptic action potential (Südhof, 1995; Rettig and Neher, 2002). Estimates of the number of primed vesicles per active zone vary from 10 for the CA1 region of the hippocampus (Schikorski and Stevens, 1997) to 130 in cat rod photoreceptors (Rao-Mirotznik et al., 1995). Early work by Llinás et al. (1976) demonstrated that a postsynaptic response can occur within 200 μs of an increase in the presynaptic Ca^{2+} current. A subsequent mathematical study indicated that this temporal restriction constrains the vesicle to be within 10's of nanometers of an open Ca^{2+} channel (Simon and Llinás, 1985). A later study of the large chick calyx synapse demonstrated that transmitter release can be evoked by the opening of a single Ca^{2+} channel, and occurs 200–400 μs after the channel opening (Stanley, 1993). Further evidence for the colocalization of channels and vesicles has been provided by studies showing that proteins forming the **core complex** that associates the vesicle with the terminal membrane, e.g. SNAP-25, syntaxin, and synaptotagmin, bind to Ca^{2+} channels in vertebrate synapses (Sheng et al., 1996, 1997; Jarvis et al., 2002). Finally, the high concentration of Ca^{2+} needed to evoke vesicle fusion, 20 μM or more (Heidelberger et al., 1994), can only be achieved if the Ca^{2+} source is located close to the Ca^{2+} acceptors.

4 Reaction-Diffusion Equations

Many models have been developed of Ca^{2+} diffusion in the presynaptic terminal or near the plasma membrane of endocrine cells (Zucker and Stockbridge, 1983; Fogelson and Zucker, 1985; Neher, 1986; Yamada and Zucker, 1992; Klingauf and Neher, 1997; Naraghi and Neher, 1997; Tang et al., 2000; Matveev et al., 2002, 2004). Most of these include mobile Ca^{2+} buffers. Examples of endogenous mobile buffers are calbindin-D28k and calretinin. Examples of exogenous mobile buffers are the Ca^{2+} chelators EGTA and BAPTA, and the fluorescent dye fura-2. The Ca^{2+}-buffer reaction diffusion equations have the form

$$\frac{\partial Ca}{\partial t} = D_c \nabla^2 Ca + R \tag{1}$$

$$\frac{\partial B}{\partial t} = D_b \nabla^2 B + R \tag{2}$$

$$\frac{\partial BC}{\partial t} = D_{bc} \nabla^2 BC - R \tag{3}$$

where Ca is the intraterminal free Ca^{2+} concentration, B is the free buffer concentration, and BC is the concentration of Ca^{2+} bound to buffer. It is assumed here that there is one mobile buffer, but others can be included in a natural way. The diffusion coefficients of the three species are D_c, D_b, and D_{bc}. The Ca^{2+} binding reaction (R) is given by

$$R = -k^+ BCa + k^- BC \tag{4}$$

where k^+ and k^- are the binding and unbinding rates, respectively. Physiological values for the diffusion coefficients and kinetic rates can be found in Allbritton et al. (1992); Pethig et al. (1989), and Klingauf and Neher (1997). If the buffer diffuses at the same rate whether Ca^{2+} is bound or not ($D_b \approx D_{bc}$), then the total concentration of buffer, $B_T = B + BC$, is described by

$$\frac{\partial B_T}{\partial t} = D_b \bigtriangledown^2 B_T \ . \tag{5}$$

From this, we see that if B_T is initially uniform, then it remains uniform. It is then possible to eliminate (2), replacing (4) with

$$R = -k^+(B_T - BC)Ca + k^- BC \ . \tag{6}$$

At the boundary (the terminal membrane), Ca^{2+} influx through an open channel is described by

$$D_c \bigtriangledown Ca = i_{Ca}/(2FA) \tag{7}$$

where i_{Ca} is the single channel Ca^{2+} current, F is Faraday's constant, and A is the channel cross-sectional area. More detailed descriptions of the reaction-diffusion equations and their numerical solution can be found in the modeling papers cited earlier and in Smith et al. (2002).

5 The Simon and Llinás Model

One of the earliest models of Ca^{2+} dynamics in the presynaptic terminal was developed by Simon and Llinás (1985). The goal of this modeling study was to determine the spatial and temporal distribution of Ca^{2+} in the squid giant synapse, a common experimental system due to its large size, during and following depolarization of the presynaptic terminal. Unlike several earlier models, the focus was on the Ca^{2+} microdomains that form at open channels.

This study produced three main predictions: (1) The opening of a Ca^{2+} channel results in a **microdomain** where the Ca^{2+} concentration is greatly elevated above the spatial average of the concentration in the terminal (the **bulk Ca^{2+} concentration**). (2) Within a microdomain, the steady-state concentration is achieved very rapidly (<1 μs) relative to the open time of the channel. (3) The macroscopic Ca^{2+} current (I_{Ca}) is not a good determinant of the microscopic distribution of Ca^{2+} in the terminal. It is the microscopic distribution that is most important for transmitter release, given the colocalization of channels and vesicles. Two other important observations were made.

First, stationary buffers have no effect on the steady-state Ca^{2+} concentration in a microdomain, but do influence the time required to reach a steady state. Second, in the vicinity of a channel the Ca^{2+} concentration is roughly proportional to i_{Ca}, the single-channel Ca^{2+} current.

The spatial extent of a microdomain depends greatly on the magnitude of the flux through the open channel, which itself depends on the voltage **driving force**, the difference between the membrane potential and the Ca^{2+} Nernst potential ($V - V_{Ca}$). When V is low (the terminal is hyperpolarized) the probability that a channel is open is small. However, the driving force is large, so the magnitude and extent of a microdomain formed at the rare open channel is large. In contrast, when V is elevated (the terminal is depolarized) the channel open probability is near one, but now the driving force is small. Hence, when V is relatively low there will be few open channels, each producing a large Ca^{2+} microdomain. When V is high there will be many open channels, each producing a small microdomain. In the latter case there could be significant overlap of microdomains, while in the former case this is unlikely to occur.

The importance of thinking in terms of the microscopic Ca^{2+} distribution, rather than in terms of the macroscopic Ca^{2+} current, is demonstrated nicely by an experiment performed in the squid giant synapse by Augustine et al. (1985). Here, the terminal was voltage clamped and the presynaptic current was recorded. The postsynaptic response, postsynaptic current, was measured simultaneously. This was done for test pulses ranging from $-33\,\mathrm{mV}$ to $57\,\mathrm{mV}$ from a holding potential of $-70\,\mathrm{mV}$ (Fig. 1). In each case, the bulk of the postsynaptic response occurred *after* the membrane potential was returned to its low holding potential. This is in spite of the fact that the time integral of the macroscopic Ca^{2+} current I_{Ca} was larger during the depolarized test pulse than following the return to the holding potential. The reason for this is that during the depolarization Ca^{2+} channels open, but the accompanying Ca^{2+} microdomains are small. When V is returned to its hyperpolarized holding potential the driving force is instantaneously increased, resulting in a rapid spike in I_{Ca}, called a **tail current** (this is not resolved in Fig. 1). Thus, before the channels have a chance to close large microdomains are formed, and these evoke the bulk of the transmitter release.

6 Steady-State Approximations for Ca^{2+} Microdomains

The colocalization of vesicles and Ca^{2+} channels suggests that transmitter release is gated by single or overlapping Ca^{2+} microdomains. This, combined with the calculations from Simon and Llinás that a steady state is achieved rapidly in a microdomain, has motivated two steady-state approximations for the Ca^{2+} concentration in a microdomain. The first, the **excess buffer approximation** (EBA), assumes that mobile buffer is present in excess and

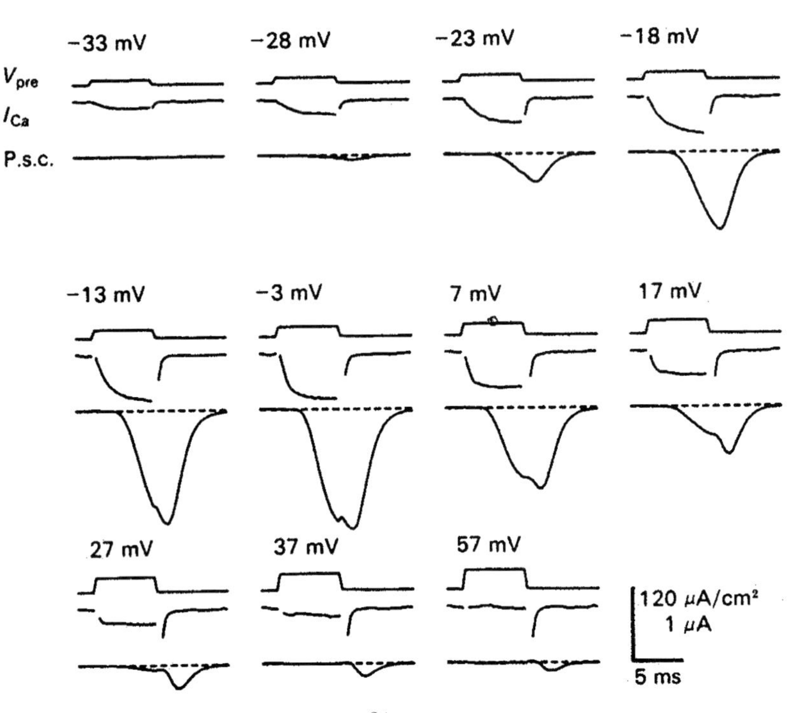

Fig. 1. Recordings of presynaptic Ca^{2+} current (I_{Ca}) and postsynaptic current (P.s.c.) at voltage-clamped terminals ($V_{\rm pre}$) of the squid giant synapse. Tail currents are not visible. Reprinted with permission from Augustine et al. (1985)

cannot be saturated. It was first derived by Neher (1986), and would be valid, for example, in the saccular hair cell where the endogenous buffer calbindin-D_{28K} is present in millimolar concentrations (Roberts, 1993), or in cells dialyzed with a high concentration of a Ca^{2+} chelator.

The steady-state EBA is:

$$Ca = \frac{\sigma}{2\pi D_c r} e^{-r/\lambda} + Ca_{bk} \tag{8}$$

where σ is the Ca^{2+} flux through an open channel, r is the distance from the channel, and Ca_{bk} is the bulk Ca^{2+} concentration. The parameter λ is the **characteristic length**, which determines the spatial extent of the microdomain. The characteristic length depends on the Ca^{2+} diffusion constant, the buffer binding rate, and the bulk free buffer concentration:

$$\lambda = \sqrt{\frac{D_c}{k^+ B_{bk}}} \, . \tag{9}$$

Figure 2 shows the graph of the EBA for two different values of λ. For the larger λ, the Ca^{2+} concentration is larger at each location and the microdomain has a greater spatial extent. The dependence of λ on D_c and k^+ (or B_{bk})

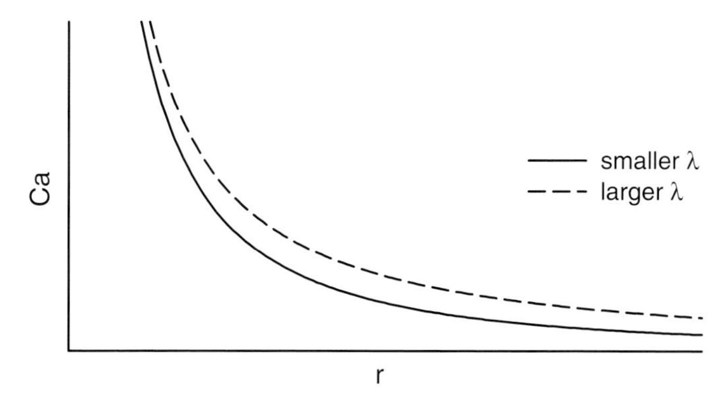

Fig. 2. The steady state Ca^{2+} concentration as a function of distance from an open Ca^{2+} channel, calculated with the EBA (8) using two different values of the characteristic length λ

is illustrated in Fig. 3. The characteristic length increases as the square root of the Ca^{2+} diffusion rate, agreeing with the intuition that the microdomain extends farther when diffusion is more rapid. In contrast, the characteristic length, and the extent of the microdomain, declines when either the buffer binding rate or the buffer concentration is increased.

A second approximation, the **rapid buffer approximation** (RBA), is based on the assumption that the Ca^{2+} buffer is fast compared to the Ca^{2+} diffusion rate (Wagner and Keizer, 1994). This leads to local equilibration, so that at each location the Ca^{2+} and buffer are in equilibrium. Thus,

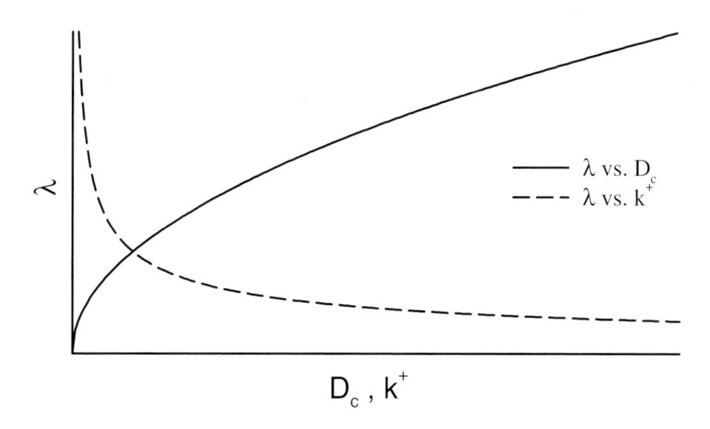

Fig. 3. The dependence of the characteristic length on the Ca^{2+} diffusion coefficient and the buffer binding rate

$$B = \frac{KB_{\rm tot}}{K + Ca} \tag{10}$$

where K is the dissociation constant ($K = k^-/k^+$) of the buffer. With this approximation, buffer may be saturated when the Ca^{2+} concentration is high, such as near an open Ca^{2+} channel, in contrast with the EBA where buffer does not saturate. Indeed, for the RBA

$$\lim_{r \to 0} B \approx 0 \tag{11}$$

while for the EBA

$$\lim_{r \to 0} B \approx B_{bk} \ . \tag{12}$$

See Smith et al. (2002) for a summary of the two approximations. An asymptotic analysis of the approximations and conditions on their validity is given in Smith et al. (2001).

The steady state RBA was derived by Smith (1996). For a single mobile buffer and a single Ca^{2+} source (channel), it is:

$$Ca = \left(-D_c K + \frac{\sigma}{2\pi r} + D_c Ca_{bk} - D_b B_{bk} + \sqrt{\Omega} \right) / 2D_c \tag{13}$$

where

$$\Omega = \left(D_c K + \frac{\sigma}{2\pi r} + D_c Ca_{bk} - D_b B_{bk} \right)^2 + 4 D_c D_b B_{\rm tot} K \ . \tag{14}$$

This approximation and the EBA were extended to multiple sources (open channels) in Bertram et al. (1999a).

7 Modeling the Postsynaptic Response

Neurotransmitter molecules released from the presynaptic terminal diffuse across the narrow synaptic cleft ($\approx$20 nm) and can bind to receptors in the postsynaptic cell. These receptors may be linked directly to an ion channel (**ionotropic receptors**) or may activate second messengers, thus providing an indirect link to ion channels (**metabotropic receptors**). We focus here on models of the action of ionotropic receptors, which are simpler than those of metabotropic receptors. For a description of both types of models see (Destexhe et al., 1994).

The postsynaptic membrane contains voltage-gated ion channels that give rise to an ionic current $I_{\rm ion}$, and ligand-gated channels that give rise to a synaptic current $I_{\rm syn}$. The postsynaptic voltage $V_{\rm post}$ changes in time according to

$$C_m \frac{dV_{\text{post}}}{dt} = -(I_{\text{ion}} + I_{\text{syn}}) \tag{15}$$

where $I_{\text{syn}} = g_{\text{syn}}(t)(V - V_{\text{syn}})$ and V_{syn} is the reversal potential for the channel (for ionotropic glutamate receptors, for example, $V_{\text{syn}} \approx 0\,\text{mV}$). The time-dependent synaptic conductance g_{syn} has been modeled in two ways: using an α-function and using a kinetic description. The **α-function** was first employed by Rall (1967) to describe the synaptic response in a passive dendrite. It is convenient to write this in terms of dimensionless time, $T = t/\tau_m$, and the dimensionless parameter $\alpha = \tau_m/t_{\text{peak}}$. Here, t is the time after the synaptic event, and t_{peak} is the time after the synaptic event at which g_{syn} reaches its peak value. The parameter $\tau_m = R_m C_m$ is the **membrane time constant** (R_m is the membrane resistance, equal to $1/g_{\text{ion}}$), which determines how rapidly the postsynaptic membrane responds to changes in current. Finally, the synaptic conductance written in terms of the α-function is:

$$g_{\text{syn}}(T) = \bar{g}_{\text{syn}} \alpha T e^{-\alpha T} \tag{16}$$

where $\bar{g}_{\text{syn}}$ is a parameter that sets the size of the response.

Synaptic conductance using (16) is shown following a synaptic event in Fig. 4. Notice the rapid rise and slow fall in conductance that is characteristic of the actual postsynaptic response. We have plotted g_{syn} for two values of the parameter α, demonstrating how doubling α affects the time course of the synaptic conductance. The conductance has been normalized to facilitate comparison of the time courses.

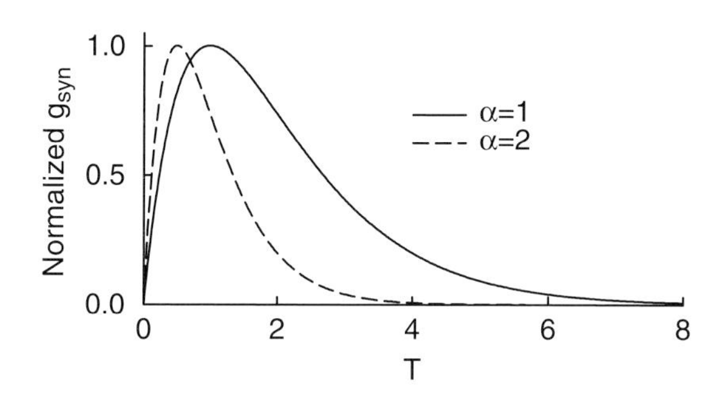

Fig. 4. Normalized synaptic conductance modeled with an α-function. T is the dimensionless time following the synaptic event. Conductance is plotted for two different values of the parameter α

The α-function approach has some disadvantages. First, it does not provide for the summation and saturation of postsynaptic responses. Second, it does not describe receptors/channels with multiple states. An alternative approach is to simulate the postsynaptic response using a kinetic model for the receptor. An excellent description of this approach is given in Destexhe et al. (1994).

The simplest receptor has two states, closed and open, and can be described by the following kinetic scheme:

$$C \underset{\beta}{\overset{\alpha(L)}{\rightleftarrows}} O$$

where C represents the closed state, O represents the open state, $\alpha(L)$ is the receptor activation rate that depends on the ligand concentration (L), and β is the receptor deactivation rate. Letting O also represent the fraction of activated receptors (or open ionotropic channels) and noting that $1 - O$ is the fraction of non-activated receptors, one can use the law of mass action to obtain:

$$\frac{dO}{dt} = \alpha(1 - O) - \beta O \ , \tag{17}$$

where α and β values for the appropriate receptor type would be used. Then, $g_{\rm syn}(t) = \bar{g}_{\rm syn} O$. One would then include (17) with the differential equations describing the postsynaptic voltage and any activation or inactivation variables. A big advantage of this approach is that receptors with complicated kinetics can be readily converted to differential equations and incorporated into the postsynaptic compartment of the model.

8 A Simple Model

In this section the release of transmitter and the postsynaptic response will be demonstrated with a simple model (Bertram, 1997). This model assumes that each releases site is 10 nm from a Ca^{2+} channel, and uses the steady-state RBA for the Ca^{2+} concentration near the channel. Because of the channel colocalization, the stochastic openings and closings of the channel would be reflected in the dynamics of transmitter release. To arrive at a deterministic model, we assume that the release site senses the **average domain Ca^{2+} concentration** rather than the stochastic Ca^{2+} signal from the colocalized channel:

$$\overline{Ca_D} = Ca_D p \tag{18}$$

where Ca_D is the microdomain Ca^{2+} concentration 10 nm from the channel and p is the probability that the channel is open. See Bertram and Sherman (1998) for a discussion of the validity of this deterministic approximation.

It is next assumed that each release site has a single low-affinity Ca^{2+} binding site (the properties of the Ca^{2+} acceptors are discussed later in more detail), with a dissociation constant of 170 µM. The differential equation for the fraction of bound Ca^{2+} acceptors, s, is then:

$$\frac{ds}{dt} = k_s^+ \overline{Ca_D}(1-s) - k_s^- s \tag{19}$$

where $k_s^+ = 0.015\ \text{ms}^{-1}\mu\text{M}^{-1}$ is the Ca^{2+} binding rate and $k_s^- = 2.5\ \text{ms}^{-1}$ is the unbinding rate.

For the postsynaptic response, we assume that the concentration of released transmitter is proportional to the probability of transmitter release (s). We then use a two-state model for the postsynaptic receptor, with fast kinetics and a reversal potential of 0 mV.

The components of neurotransmission simulated with this model are shown in Fig. 5. A presynaptic action potential is generated that peaks at the time indicated by the dashed line (Fig. 5A). During the downstroke of the impulse the Ca^{2+} microdomains are largest (similar to the large tail currents described earlier), so the average domain Ca^{2+} concentration peaks shortly after the peak of the impulse (Fig. 5B). The rise in domain Ca^{2+} causes a spike in the release probability (Fig. 5C), which in turn results in a relatively rapid rise and slow decay in the postsynaptic membrane potential (Fig. 5D). The rapid rise is due to I_{syn}, which is activated during the spike in release probability. The slow decay in V_{post} reflects the membrane properties of the postsynaptic compartment. Thus, we see that the postsynaptic response occurs several milliseconds after the presynaptic impulse, consistent with experimental observations. Also, the postsynaptic potential outlasts the presynaptic signal that initiated it. This allows the postsynaptic potential to summate during a train of presynaptic impulses (Fig. 6).

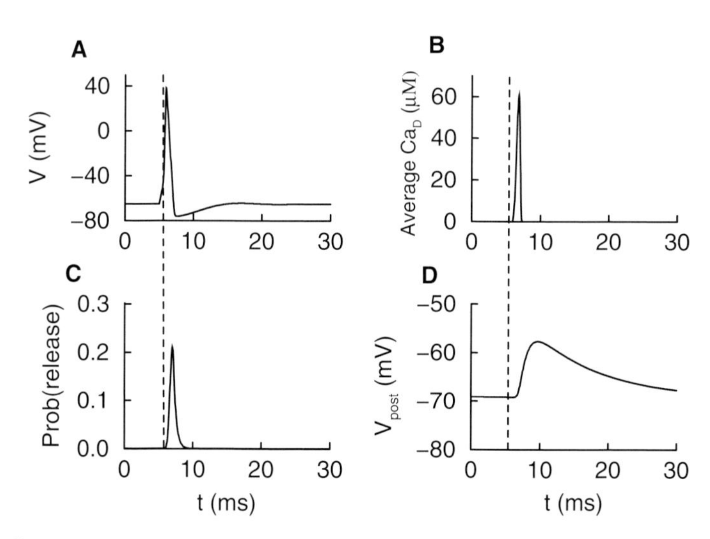

Fig. 5. Simulation of transmitter release and a postsynaptic response. The postsynaptic response is delayed several milliseconds from the presynaptic impulse, and it is longer lasting

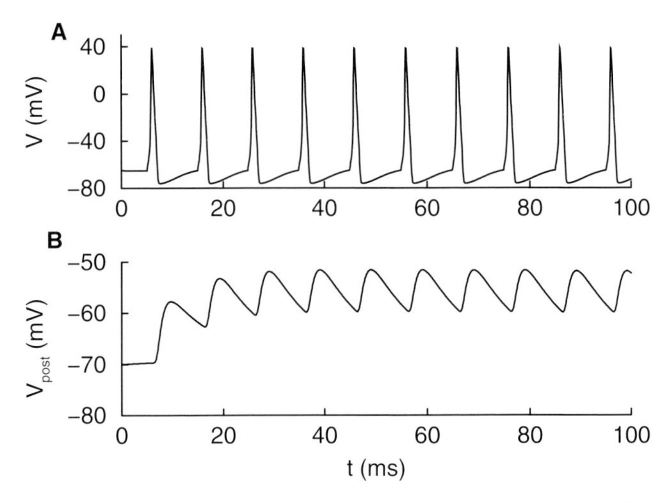

Fig. 6. Simulated postsynaptic response to a presynaptic train of impulses. The postsynaptic membrane time constant and the stimulus frequency determine the degree of summation in the postsynaptic response

9 Short-Term Plasticity

The synapse is highly plastic, exhibiting several forms of short-term and long-term plasticity. Long-term plasticity occurs in both presynaptic and postsynaptic regions and typically involves insertion of receptors or channels into the membrane and/or targeted gene expression. Short-term plasticity is often accomplished without any structural modifications to the synapse, and can be due to things such as Ca^{2+} accumulation, activation of second messengers, and phosphorylation (Zucker and Regehr, 2002). Our focus will be on short-term plasticity.

Short-term plasticity is often measured using a paired-pulse protocol. Here, the presynaptic neuron is stimulated twice, with an interstimulus interval of δt. The postsynaptic response, either current or potential, is recorded for each stimulus, and the ratio of the second response to the first is the measure of plasticity, $SP_2 = P_2/P_1$. If $SP_2 > 1$ then the second response is facilitated; if $SP_2 < 1$ then the second response is depressed. The duration of the facilitation or depression is determined by increasing δt until $SP_2 \approx 1$.

Another protocol used to measure plasticity is to apply an impulse train (often called a "tetanus") to the presynaptic neuron and record the postsynaptic response throughout the train. Then $SP_n = P_n/P_1$ provides a measure of the enhancement or depression of the nth response relative to the first. It is not uncommon for SP_n to indicate enhancement at the beginning of the train and depression later in the train, as various plasticity mechanisms compete with different time scales. A variation of this stimulus protocol is to induce a "conditioning" impulse train and then, at a time δt following the train, to induce a single impulse, the "test stimulus". The ratio of the postsynaptic responses to the test stimulus with and without the conditioning train then provides a measure of plasticity induced by the train. Often δt is varied to determine the time constant of decay of the plasticity.

Synaptic enhancement occurs at several time scales. **Facilitation** occurs at the fastest time scale, and is often measured using a paired-pulse protocol (Fig. 13). This type of enhancement is often subdivided into F1 facilitation, which lasts tens of milliseconds, and F2 facilitation, which lasts hundreds of milliseconds. **Augmentation** develops during an impulse train, and grows and decays with a time constant of 5–10 sec. Finally, **post-tetanic potentiation** is like augmentation in that it is induced by an impulse train, however it is longer-lasting, with a time constant of ≈ 30 sec. The distinct time constants for the various forms of enhancement suggest that they are produced by different mechanisms.

Synaptic **depression** is observed in most central nervous system synapses and less frequently in neuromuscular junctions. In central synapses, the postsynaptic response typically declines throughout a train of presynaptic stimuli (Fig. 7), or first rises and then declines later in the train. In many cases the degree of depression increases with the stimulus frequency (Abbott et al., 1997; Tsodyks and Markram, 1997), although in some cases the opposite is true (Shen and Horn, 1996). This difference in the frequency dependence of depression suggests at least two mechanisms for this form of plasticity.

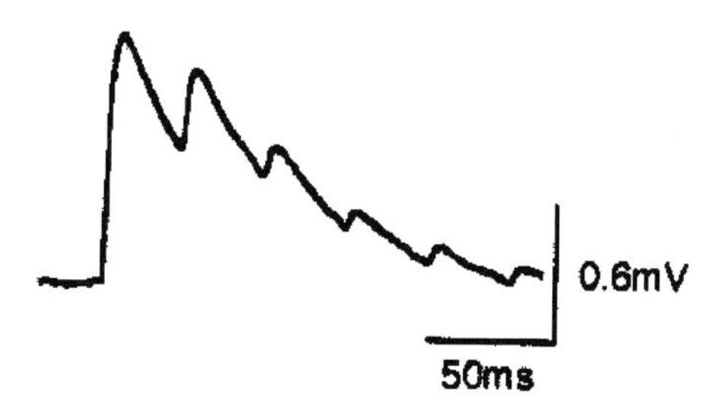

Fig. 7. The postsynaptic response depresses throughout a train of presynaptic impulses in this pyramidal neuron from the rat cortex. Reprinted with permission from Markram et al. (1998a)

10 Mathematical Models for Facilitation

The mechanism for facilitation of transmitter release has been debated for more than three decades, ever since Katz and Miledi (1968) and Rahamimoff (1968) suggested that facilitation is due to the buildup of Ca^{2+} bound to release sites. It has been clear from the beginning that Ca^{2+} plays a key role, but the difficulties inherent in measuring Ca^{2+} microdomains, and the small size of the synapse, has obscured the mechanism by which Ca^{2+} produces facilitation. Thus, mathematical models of various facilitation mechanisms have been developed, and predictions from these models compared with experiments. Three main mechanisms for facilitation have been postulated: (1) residual free Ca^{2+}, (2) residual bound Ca^{2+}, and (3) saturation of endogenous Ca^{2+} buffers. Models for each of these will be discussed below, beginning with a series of models developed by Zucker and colleagues for a residual free Ca^{2+} mechanism of facilitation.

11 Residual Free Ca^{2+} Models

The Fogelson-Zucker model (Fogelson and Zucker, 1985) was developed to describe transmitter release and its facilitation in the squid giant synapse. The authors first considered a model in which Ca^{2+} diffuses in one dimension. Cylindrical coordinates were used, and it was assumed that Ca^{2+} enters through the whole surface membrane of the terminal and diffuses only in the radial direction. The model included stationary buffers. The authors also considered a 3-dimensional Ca^{2+} diffusion model where Ca^{2+} enters through an array of channels in the terminal membrane. Each channel was considered to be a point source on the boundary. Transmitter release was modeled as a power of the free Ca^{2+} concentration in the terminal, $R = kCa_{bk}^{n}$. The stoichiometry $n = 5$ was chosen to best fit experimental data on facilitation in the squid synapse.

The main result of this work was that during a train of impulses the free Ca^{2+} concentration in the terminal could accumulate, and could potentially account for the facilitation in transmitter release observed in the squid synapse. That is, since $R = kCa_{bk}^{5}$ and since Ca_{bk} builds up during the impulse train, the release from each impulse is facilitated relative to that from the previous impulse. Importantly, this model predicts that it is the diffusion of Ca^{2+} that determines both transmitter release and its facilitation, and that other factors such as the kinetics of Ca^{2+} binding to or unbinding from the release sites, or the effects of mobile buffers, are secondary.

The Fogelson-Zucker model motivated many experimental studies aimed at testing this residual free Ca^{2+} mechanism for facilitation. Parnas and colleagues pointed out that the Fogelson-Zucker model failed to account for several experimental findings (Parnas et al., 1989). In particular, facilitation has been shown to be very dependent on temperature, while the diffusion of Ca^{2+} would only be weakly temperature dependent. They also pointed to some early experiments by Datyner and Gage, who studied the shape of the release time course during facilitation and with different concentrations of external Ca^{2+} by normalizing the peaks of the responses (Datyner and Gage, 1980). Inducing a train of three presynaptic impulses at 65 Hz, they found that although the response clearly facilitated, the normalized time course was invariant throughout the train (Fig. 8). Similarly, they found time course invariance

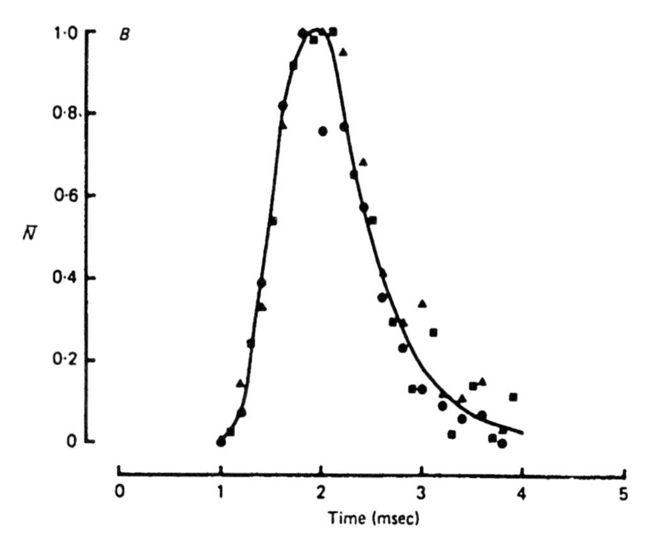

Fig. 8. The normalized release time course is invariant during a 65 Hz train of three stimuli that produces facilitation. Responses to the first (*circles*), second (*triangles*), and third stimulus (*squares*) normalized to have the same peak values. Reprinted with permission from Datyner and Gage (1980)

when comparing the response in low external Ca^{2+} and high external Ca^{2+} (the latter produced a larger peak response). Parnas and colleagues pointed out that the Fogelson-Zucker model did not exhibit time course invariance. A final criticism of the model was that the colocalization of low-affinity Ca^{2+} acceptors and Ca^{2+} channels makes it unlikely that a small (sub-micromolar) buildup in the Ca^{2+} concentration could have a significant effect on the magnitude of Ca^{2+} release.

A second-generation residual free Ca^{2+} model that addressed these problems was developed by Yamada and Zucker (1992). As in the earlier model, intraterminal Ca^{2+} diffused in three dimensions. Now, however, the Ca^{2+} binding kinetics at Ca^{2+} acceptors on the release sites were included. The authors assumed that there are four identical low-affinity acceptors (denoted by X) that act as triggers to secretion, and a high-affinity site (denoted by Y) that is responsible for facilitation. The kinetic relations are:

$$X + Ca \underset{k_x^-}{\overset{4k_x^+}{\rightleftarrows}} CaX + Ca \underset{2k_x^-}{\overset{3k_x^+}{\rightleftarrows}} Ca_2X + Ca \underset{3k_x^-}{\overset{2k_x^+}{\rightleftarrows}} Ca_3X + Ca \underset{4k_x^-}{\overset{k_x^+}{\rightleftarrows}} Ca_4X$$

and

$$Y + Ca \underset{k_y^-}{\overset{k_y^+}{\rightleftarrows}} CaY.$$

Release occurs when all four X sites are bound and the Y site is bound,

$$Ca_4X + CaY \underset{k_r^-}{\overset{k_r^+}{\rightleftharpoons}} R$$

The binding kinetics are converted to differential equations using the law of mass action.

As in the Fogelson-Zucker model, most of the facilitation in this model is due to the buildup of free Ca^{2+}. During a train of impulses Ca^{2+} accumulates and binds to the high-affinity Y sites, which bind and unbind Ca^{2+} slowly. These slow rates average the fast Ca^{2+} signals generated by impulses during an impulse train, and make this a hybrid residual free Ca^{2+}/residual bound Ca^{2+} model. The temperature dependence of facilitation is easily explained as a temperature-dependent effect on Ca^{2+} binding to the Y site. In addition, the normalized release time course is relatively invariant during a facilitating train and when the external Ca^{2+} concentration is changed.

Following publication of the Yamada-Zucker model a key experiment was performed showing that facilitation, augmentation, and post-tetanic potentiation were all reduced following activation of the photolabile Ca^{2+} chelator diazo-2 (Kamiya and Zucker, 1994). This chelator has low Ca^{2+} affinity unless photolyzed by UV light, which greatly increases the affinity. Diazo-2 was injected iontophoretically into an isolated crayfish motor neuron. Following injection, a train of presynaptic impulses was induced and the end junction potential (e.j.p.) recorded. The response was clearly facilitated by the end of the 10-impulse train. This facilitation was reduced by half (but not eliminated) when a UV light flash was applied prior to the test impulse at the end of the train (Fig. 9). Since UV light increases the Ca^{2+} affinity of diazo-2 to a value similar to that of BAPTA, the natural conclusion was that the photoactivated diazo-2 buffered down the residual free Ca^{2+} and thereby reduced the facilitation. Additional experiments using the Ca^{2+} chelators BAPTA and EGTA generally support the hypothesis that residual free Ca^{2+} is a key element of facilitation and other forms of short-term enhancement (Atluri and Regehr, 1996; Fischer et al., 1997a). It should be noted, however, that some labs have found that Ca^{2+} chelators have no effect on facilitation (Robitaille and Charlton, 1991; Winslow et al., 1994).

The rapid rate at which facilitation was reduced by activated diazo-2 was at odds with the slow-unbinding facilitation site postulated by Yamada and Zucker. However, if the unbinding rate were increased to satisfy the Kamiya-Zucker experiment, then the dynamics of the Y site would be determined more by the opening and closing of the colocalized channel than by the residual free Ca^{2+}. For this reason, the Zucker group formulated a new model of transmitter release and facilitation in which the high-affinity facilitation site is located away from a Ca^{2+} channel (Tang et al., 2000). In this model, three low-affinity X sites are postulated, at a distance of 10–20 nm from a Ca^{2+} channel. The high-affinity Y site is postulated to lie 80–100 nm from the nearest Ca^{2+} channel. The kinetics are basically the same as in the Yamada-Zucker model,

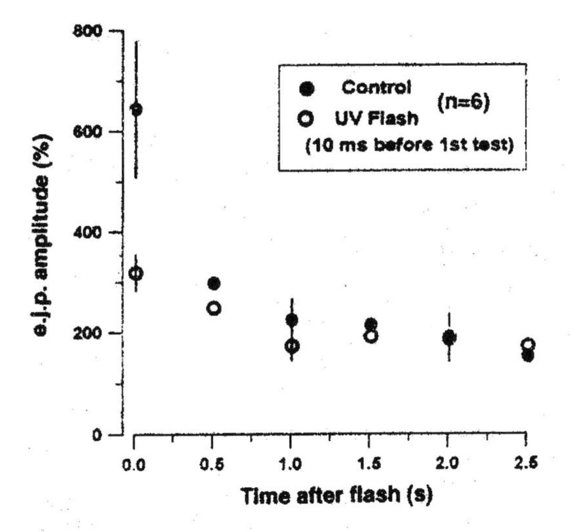

Fig. 9. Facilitation following a 50 Hz train of 10 impulses was reduced following UV activation of the photolabile Ca^{2+} chelator diazo-2. Data from a crayfish motor neuron. Reprinted with permission from Kamiya and Zucker (1994)

except that the Y site is now assumed to unbind Ca^{2+} rapidly, allowing the model to reproduce the Kamiya-Zucker experiment.

The Tang et al. model also aimed to reproduce the authors' experiments showing that both the accumulation and the decay of facilitation are affected by the fast high-affinity Ca^{2+} buffer fura-2 and that facilitation grows supralinearly during an impulse train (Tang et al., 2000). Unfortunately, the computer software developed for the model had several errors, and although the model originally seemed to reproduce most of the data, once the errors were corrected the fit to the data was not good (Matveev et al., 2002). However, once the corrected model was recalibrated, it was able to reproduce most of the data, but not all, from Tang et al. Significantly, the model was able to produce supralinear faciliation, and the facilitation was reduced when a high-affinity buffer was simulated (Fig. 10). The success of the model simulations required several stringent assumptions. It was necessary to assume that the high-affinity buffer fura-2 is immobilized and the Ca^{2+} diffusion coefficient is reduced fivefold near the active zone, possibly due to tortuosity. It was also necessary to reduce the diffusion coefficient of fura-2 100-fold in the rest of the terminal (Matveev et al., 2002). Since these assumptions are questionable, it would seem that while this latest implementation of the residual free Ca^{2+} hypothesis captures some features of facilitation, significant modifications are needed to fully explain the phenomenon.

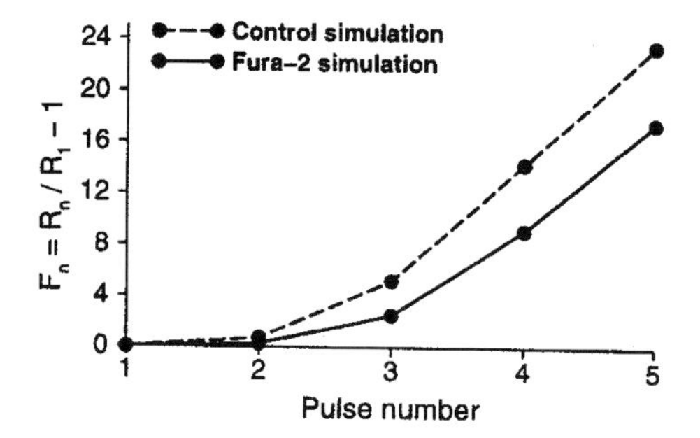

Fig. 10. In the model of Matveev et al. (2002), faciliation grows supralinearly during a pulse train. The degree of facilitation is reduced by simulated application of the high-affinity buffer fura-2. Reprinted with permission from Matveev et al. (2002)

12 A Residual Bound Ca^{2+} Model for Facilitation

The mechanism originally proposed by Katz and Miledi (1968) and Rahamimoff (1968), that facilitation is due to the accumulation of Ca^{2+} bound to release sites, fell out of favor as models for a residual free Ca^{2+} mechanism established the dominant paradigm. This paradigm was challenged by an alternate hypothesis of Stanley (Stanley, 1986) that was implemented as a mathematical model by Bertram et al. (1996). This was based on experimental data from the squid giant synapse suggesting that the Ca^{2+} cooperativity of transmitter release decreased during facilitation, and that the steady-state facilitation increased in a step-like fashion with the stimulus frequency (Stanley, 1986). Bertram et al. showed that both of these phenomena could be explained if one assumed that each release site has four Ca^{2+} acceptors with differing Ca^{2+} affinities and unbinding rates.

In this model, each Ca^{2+} acceptor can be described by a first order kinetic scheme,

$$U_j + Ca \underset{k_j^-}{\overset{k_j^+}{\rightleftharpoons}} B_j$$

where U_j represents an unbound acceptor ($j = 1\text{–}4$, for the four acceptors) and B_j represents a Ca^{2+}-bound acceptor. Release occurs when all four acceptors are occupied,

$$R = B_1 B_2 B_3 B_4 \, . \tag{20}$$

Binding and unbinding rates, k_j^+ and k_j^- respectively, were chosen to produce steps in the frequency-dependence of facilitation. These gave unbinding time constants ($1/k_j^-$) of 2.5 s, 1 s, 10 ms, and 0.1 ms, and dissociation constants (k_j^-/k_j^+) of 108 nM, 400 nM, 200 μM, and 1,334 μM for the four acceptors (denoted S_1–S_4).

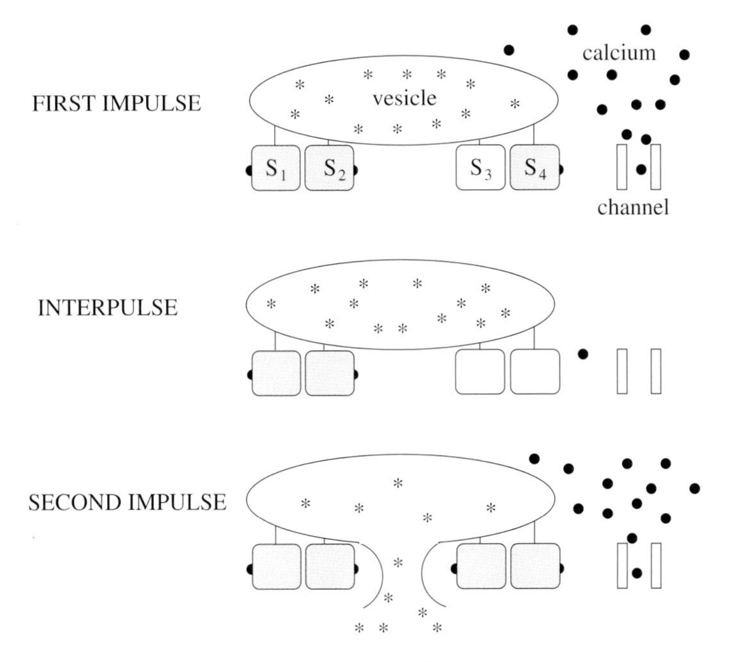

Fig. 11. Illustration of the residual bound Ca^{2+} model of facilitation by Bertram et al. (1996). Each box represents a Ca^{2+} acceptor. Reprinted from Bertram and Sherman (1998)

The mechanism by which step-like facilitation is produced with this model is illustrated in Fig. 11. Prior to the first presynaptic impulse none of the acceptors are occupied. With the first impulse, a colocalized Ca^{2+} channel opens, and Ca^{2+} entering through the channel forms a microdomain. In this illustration, Ca^{2+} within the microdomain binds to three of the four acceptors. Since one acceptor remains unbound the vesicle does not fuse with the membrane. After the first impulse Ca^{2+} unbinds from the acceptor with the smallest unbinding time constant (site S_4). However, if the second impulse arrives before Ca^{2+} can unbind from sites S_1 and S_2 then Ca^{2+} from this second impulse must only bind to two of the four acceptors to induce vesicle fusion and transmitter release. In this way, using a residual bound Ca^{2+} mechanism, the release probability is facilitated. The reduced number of sites needed to gate facilitated release accounts for the drop in the Ca^{2+} cooperativity reported by Stanley, and the three distinct slow unbinding rates account for the three steps in the frequency-dependence of facilitation. Because facilitation in this model is due to slow Ca^{2+} unbinding rather than buildup of free Ca^{2+}, the mechanism would be highly temperature dependent. Also, simulations and mathematical analysis showed that the release time course exhibits the required invariance properties (Bertram et al., 1996; Bertram and Sherman, 1998). However, this model does not account for the reduction of facilitation caused by Ca^{2+} chelators. In particular, it does not reproduce the results of the Kamiya-Zucker diazo-2 experiment, unless modified to include effects of residual free Ca^{2+} as well as residual bound Ca^{2+}.

13 A Model for Facilitation Based on Buffer Saturation

Another mechanism for facilitation was suggested by simulations of Ca^{2+} and buffer diffusion near a membrane (Klingauf and Neher, 1997; Neher, 1998). In this model, Ca^{2+} buffer saturates during an impulse train, so that the amount of free Ca^{2+} introduced near the membrane by an impulse increases throughout the train, producing facilitated transmitter release.

The conditions in which buffer can saturate and produce facilitation were clarified in a study by Matveev et al. (2004). They found that robust facilitation can be achieved either by a global saturation of a highly mobile buffer (like fura-2 or BAPTA) in the entire presynaptic terminal, or saturation of an immobile buffer local to the release sites. Figure 12 shows how facilitation occurs with this mechanism. Here a 100 Hz train of equal Ca^{2+} current pulses is applied to the model terminal. A mobile buffer is present at a concentration of 500 μM (free buffer concentration is plotted in the bottom panel). The free Ca^{2+} concentration a distance of 60 nM from a cluster of channels is shown in the middle panel. During the first pulse Ca^{2+} concentration rises and binds

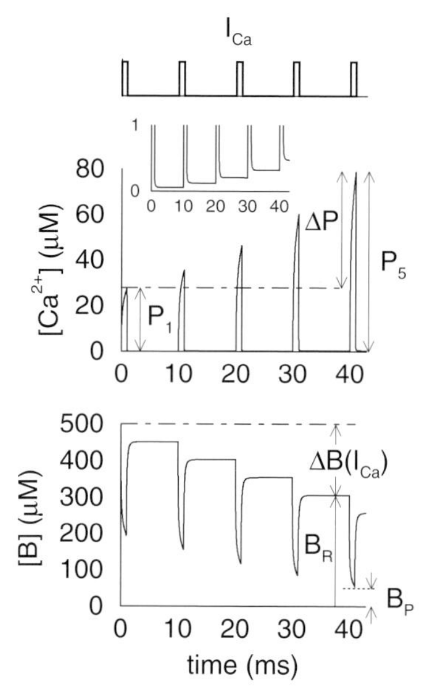

Fig. 12. In this simulation by Matveev et al. (2004) the free Ca^{2+} concentration rises to a higher level with each successive pulse of Ca^{2+} current, due to the saturation of mobile buffer. Reprinted with permission from Matveev et al. (2004)

to buffer, transiently reducing the free buffer concentration. Between the first and second pulse much of the Ca^{2+} is extruded from the terminal, but some remains in equilibrium with the buffer. This residual binding to buffer decreases the amount of buffer available to chelate Ca^{2+} during the second pulse, so the free Ca^{2+} concentration rises to a higher level during the second pulse (middle panel), resulting in facilitated transmitter release.

There is recent experimental evidence supporting the buffer saturation mechanism for facilitation. Blatow et al. (2003) examined terminals of GABA-ergic interneurons in the mouse neocortex and mouse hippocampal mossy fiber terminals. Both of these contain the endogenous fast Ca^{2+} buffer calbindin-D28k (CB), and facilitation is robust. In the interneuron synapses, CB knock-out eliminated the facilitation. However, facilitation was rescued by addition of the exogenous buffer BAPTA. That is, in these terminals addition of a high-affinity buffer increased facilitation, contrary to what would be expected if facilitation were due to either residual free or bound Ca^{2+}. Another property of the buffer saturation mechanism is that facilitation should increase when the external Ca^{2+} concentration is raised, since the resulting greater Ca^{2+} influx would produce more buffer saturation. In other models of facilitation this maneuver would decrease facilitation (Zucker, 1989). Blatow et al. showed that in the CB-containing mossy fiber terminals the facilitation increased when external Ca^{2+} concentration was raised (Fig. 13). Interestingly, they also showed that a second type of excitatory synapse in the hippocampus, the Schaffer collateral to CA1 pyramidal cell synapse, has properties consistent with a residual free Ca^{2+} mechanism for facilitation. Further, in CB knockouts in mossy fiber synapses, facilitation was reduced but not lost. The residual facilitation had all the properties expected from a residual free Ca^{2+} mechanism (Fig. 13).

The study by Blatow et al. (2003) not only provides strong evidence for the buffer saturation mechanism for facilitation, but it also serves as an excellent demonstration that different facilitation mechanisms may exist in different synapses, and that facilitation in some synapses may be due to a combination of mechanisms.

14 Synaptic Depression

Use-dependent depression of transmitter release is common in central synapses. This is reflected in a reduction in the magnitude of the postsynaptic current elicited by action potentials in an impulse train (Fig. 7). One mechanism for this is the partial depletion of the readily releasable pool of vesicles (Zucker and Regehr, 2002). Simply put, prior impulses reduce the number of primed vesicles that can be released by subsequent impulses. The recovery time from this form of depression is determined by the time required to refill the readily releasable pool.

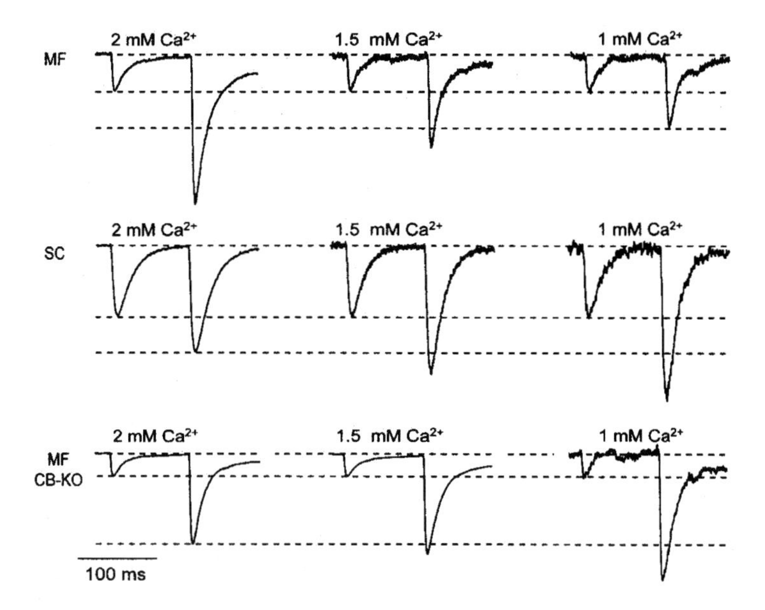

Fig. 13. In mossy fiber (MF) terminals containing CB, paired pulse facilitation becomes larger as the external Ca^{2+} concentration is increased. In Schaffer collateral (SC) synapses, facilitation decreases. In MF terminals with CB knocked out, the effect of external Ca^{2+} on facilitation is the same as in SC terminals. Reprinted with permission from Blatow et al. (2003)

An interesting link between depression and information processing was made by Abbott et al. (1997) and Tsodyks and Markram (1997). Cortical and neocortical neurons receive synaptic input from thousands of afferents, which fire at rates ranging from a few Hz to several hundred Hz. If synapses responded linearly to the presynaptic impulse frequency, then high-frequency inputs would dominate low-frequency inputs. Depression prevents this from happening by acting as a gain control, reducing the responses to impulses in high-frequency trains proportionally to the frequency of the train. That is, if $A(r)$ denotes the steady state postsynaptic response amplitude to an impulse in a train of frequency r, then for r sufficiently large $A(r) \propto 1/r$ in cortical and neocortical synapses. Thus, the total synaptic conductance during one second of stimulation, $rA(r)$, is approximately constant at all frequencies above some threshold (Fig. 14). In this way, low- and high- frequency trains are equalized.

Suppose that an impulse train is applied at frequency r and the system reaches a steady state. Now suppose that the frequency is suddenly changed by Δr. Before the system has established its new steady state, the change in the synaptic response (ΔR) is

$$\Delta R = \Delta r A(r) \approx \Delta r / r \ . \tag{21}$$

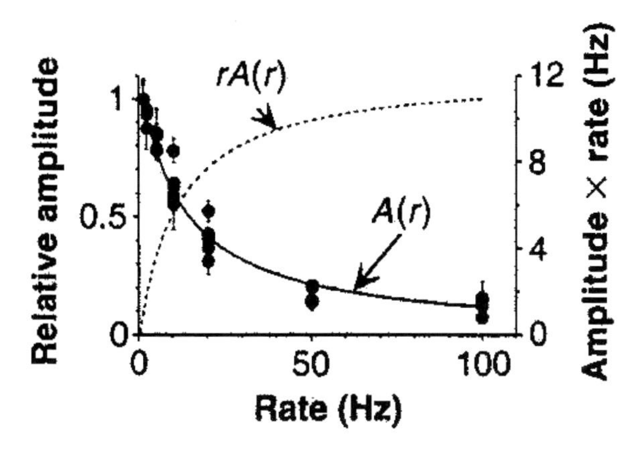

Fig. 14. In synapses of cortical neurons, the steady state postsynaptic response amplitude $A(r)$ has a $1/r$ dependence on the stimulus frequency r. The total synaptic response per second of stimulus, $rA(r)$, is roughly constant above a threshold frequency. Reprinted with permission from Abbott et al. (1997)

Thus, an increase from 10 Hz to 11 Hz ($\Delta r/r = 0.1$) will have the same initial effect as an increase from 100 Hz to 110 Hz ($\Delta r/r = 0.1$). That is, the change in the initial response is proportional to the **relative change** in the input frequency (Abbott et al., 1997; Tsodyks and Markram, 1997). The ability of neurons to sense relative changes in stimuli is a highly desirable feature, and it is remarkable that a mechanism for this is something as simple as depletion of resources.

15 G Protein Inhibition of Presynaptic Ca^{2+} Channels

Depletion of the readily releasable vesicle pool is not the only mechanism for synaptic depression. Another common mechanism is through the action of G proteins. The presynaptic terminal contains various G protein-coupled receptors. These receptors are activated by a variety of ligands, including the neurotransmitters GABA, adenosine, glutamate, dopamine and serotonin (Hille, 1994). Binding of a ligand molecule to a receptor activates the associated G protein. The G protein is a heterotrimer, with α, β, and γ subunits. When activated, GTP replaces GDP bound to the G protein, and the Gα subunit dissociates from the G$\beta\gamma$ dimer, which remains tethered to the membrane. Once activated in this way, G$\beta\gamma$ can bind to Ca^{2+} channels, putting them into a **reluctant state**, with a decreased opening rate and an increased closing rate (Ikeda, 1996).

The agonist for the G protein-coupled receptor can be a hormone, or it can be released from the postsynaptic neuron, or from the presynaptic terminal itself (Fig. 15). In fact, many synapses have G protein-coupled autoreceptors that are specific for the transmitter released from the terminal (Wu and

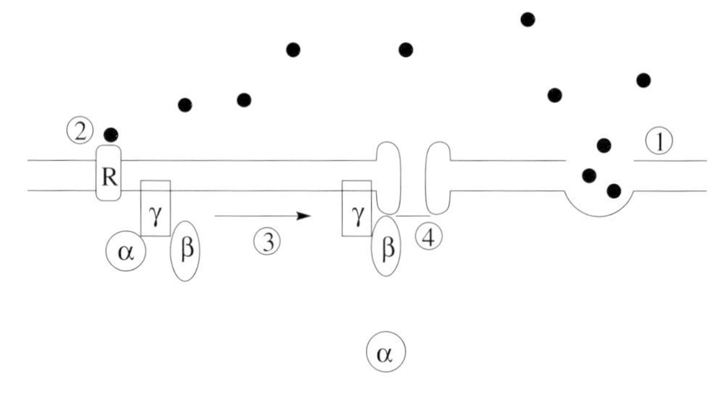

Fig. 15. Transmitter released from the presynaptic terminal (1) can bind to a G protein-coupled autoreceptor (2), activating the associated G protein (3). The $G\beta\gamma$ dimer can then bind to a Ca^{2+} channel (4), putting it into a reluctant state

Saggau, 1997; Chen and van den Pol, 1998) and G protein-mediated autoinhibition has been demonstrated under physiological conditions (Benoit-Marand et al., 2001).

An interesting feature of G protein inhibition of Ca^{2+} channels is that in most cases the inhibition is relieved by membrane depolarization. This is due to the unbinding of $G\beta\gamma$ from the Ca^{2+} channel (Zamponi and Snutch, 1998). Thus, depolarization of the terminal has two opposing effects. One is to elicit transmitter release, which results in activation of G proteins through autoreceptor binding. The increased G protein activation tends to inhibit further transmitter release. The second effect of depolarization is to cause activated G proteins to unbind from the Ca^{2+} channels, relieving the inhibition. Thus, the overall effect of the G protein pathway will depend upon which of these two opposing actions dominates.

A mathematical model for the action of G proteins on Ca^{2+} channels was developed by Boland and Bean (1993). This model was subsequently simplified and coupled to a model of transmitter release and autoreceptor binding (Bertram and Behan, 1999). Through numerical simulations, it was shown that G protein autoinhibition acts as a high-pass filter, allowing high-frequency trains of impulses to pass from the presynaptic to the postsynaptic cell, while low-frequency trains are filtered out (Bertram, 2001). The high-frequency trains are transmitted since the mean membrane potential of the synapse is elevated during such trains, relieving the G protein inhibition. Low-frequency trains result in less transmitter release, and thus a lower level of G protein activation, but those G proteins that are activated are much more effective at inhibiting Ca^{2+} channels and subsequent transmitter release.

One role of the G protein-induced high-pass filtering could be to remove noise, in the form of low-frequency signals, from a neural system. For example, consider a neural network in which a 5-by-5 input layer of neurons innervates a 5-by-5 output layer. Suppose that the G protein pathway operates in each of the input layer synapses. Finally, suppose that a signal in the input layer is degraded by low-frequency noise (Fig. 16A). Here, the signal is a spatial pattern (an X) consisting of high-frequency impulse trains in 5 of the input layer neurons. Other input-layer neurons produce impulse trains at various lower frequencies (color and size coded in Fig. 16). Because of the filtering action of G proteins, the signal is transmitted to the postsynaptic cells, while the noise is attenuated (Fig. 16B). Thus, the G protein pathway effectively increases the spatial contrast of the "image".

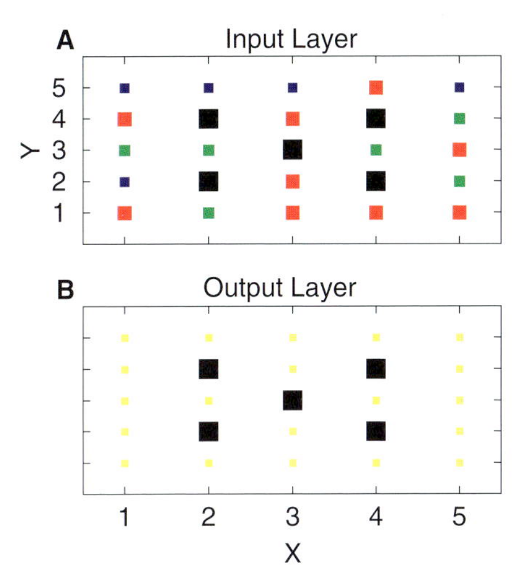

Fig. 16. G protein-mediated high-pass filtering can increase the spatial contrast of an image by increasing the signal-to-noise ratio. **(A)** A 5×5 grid of input cells fires at various frequencies r (Black, $r > 100$ Hz; red, $70 < r \leq 100$; green, $40 < r \leq 70$; blue, $10 < r \leq 40$; yellow, $r < 10$). **(B)** Each cell in the 5×5 output layer recieves synaptic input from one cell in the input layer. As a result of autoinhibition of the presynaptic transmitter release, the signal-to-noise ratio in the output layer is much greater than that in the input layer. Reprinted from Bertram et al. (2002)

16 Conclusion

Mathematical modeling and computer simulations have played a large role in unveiling the mechanisms of synaptic transmitter release and its short term plasticity. Yet, while progress has been made, there is much that remains unclear. For example, none of the models of facilitation can account for all of the key features of the experimental data from the crayfish neuromuscular junction. How should these models be modified or combined to better account for the data? What is the best way to model Ca^{2+} in the terminal? Solving 3-dimensional reaction diffusion equations is the most accurate method, but is the simpler approach of using steady state approximations sufficient in many cases? The answer to this will certainly depend on the extend of overlap of Ca^{2+} microdomains at the release sites, which may vary greatly from synapse to synapse. It appears that there is still a great deal of modeling and analysis of transmitter release and its plasticity that can be done, in parallel with the many experimental studies that are being performed with ever-improving tools.

Acknowledgements

The author thanks the National Science Foundation for financial support through grant DMS-0311856.

References

L. F. Abbott, J. A. Varela, K. Sen, and S. B. Nelson, *Synaptic depression and cortical gain control*, Science, 275 (1997), pp. 220–224.

S. Aharon, H. Parnas, and I. Parnas, *The magnitude and significance of Ca^{2+} domains for release of neurotransmitter*, Bull. Math. Biol., 56 (1994), pp. 1095–1119.

N. L. Allbritton, T. Meyer, and L. Stryer, *Range of messenger action of calcium ion and inositol 1,4,5-trisphophate*, Science, 258 (1992), pp. 1812–1815.

P. P. Atluri and W. G. Regehr, *Determinants of the time course of facilitation at the granule cell to Purkinje cell synapse*, J. Neurosci., 16 (1996), pp. 5661–5671.

G. J. Augustine, M. P. Charlton, and S. J. Smith, *Calcium entry and transmitter release at voltage-clamped nerve terminals of squid*, J. Physiol. (Lond.), 367 (1985), pp. 163–181.

M. Benoit-Marand, E. Borrelli, and F. Gonon, *Inhibition of dopamine release via presynaptic D2 receptors: Time course and functional characteristics* in vivo, J. Neurosci., 21 (2001), pp. 9134–9141.

R. Bertram, *A simple model of transmitter release and facilitation*, Neural Comput., 9 (1997), pp. 515–523.

——, *Differential filtering of two presynaptic depression mechanisms*, Neural Comput., 13 (2001), pp. 69–85.

R. Bertram, M. I. Arnot, and G. W. Zamponi, *Role for G protein $G\beta\gamma$ isoform specificity in synaptic signal processing: A computational study*, J. Neurophysiol., 87 (2002), pp. 2612–2623.

R. Bertram and M. Behan, *Implications of G-protein-mediated Ca^{2+} channel inhibition for neurotransmitter release and facilitation*, J. Comput. Neurosci., 7 (1999), pp. 197–211.

R. Bertram and A. Sherman, *Population dynamics of synaptic release sites*, SIAM J. Appl. Math., 58 (1998), pp. 142–169.

R. Bertram, A. Sherman, and E. F. Stanley, *Single-domain/bound calcium hypothesis of transmitter release and facilitation*, J. Neurophysiol., 75 (1996), pp. 1919–1931.

R. Bertram, G. D. Smith, and A. Sherman, *A modeling study of the effects of overlapping Ca^{2+} microdomains on neurotransmitter release*, Biophys. J., 76 (1999), pp. 735–750.

M. Blatow, A. Caputi, N. Burnashev, H. Monyer, and A. Rozov, *Ca^{2+} buffer saturation underlies paired pulse facilitation in calbindin-D28k-containing terminals*, Neuron, 38 (2003), pp. 79–88.

L. M. Boland and B. P. Bean, *Modulation of N-type calcium channels in bullfrog sympathetic neurons by luteinizing hormone-releasing hormone: kinetics and voltage dependence*, J. Neurosci., 13 (1993), pp. 516–533.

G. Chen and A. N. van den Pol, *Presynaptic $GABA_B$ autoreceptor modulation of P/Q-type calcium channels and GABA release in rat suprachiasmatic nucleus neurons*, J. Neurosci., 18 (1998), pp. 1913–1922.

N. B. Datyner and P. W. Gage, *Phasic secretion of acetylcholine at a mammalian neuromuscular junction*, J. Physiol. (Lond.), 303 (1980), pp. 299–314.

A. Destexhe, Z. F. Mainen, and T. J. Sejnowski, *Synthesis of models for excitable membranes, synaptic transmission and neuromodulation using a common kinetic formalism*, J. Comput. Neurosci., 1 (1994), pp. 195–230.

F. A. Dodge, Jr. and R. Rahamimoff, *Co-operative action of calcium ions in transmitter release at the neuromuscular junction*, J. Physiol. Lond, 193 (1967), pp. 419–432.

T. M. Fischer, R. S. Zucker, and T. J. Carew, *Activity-dependent potentiation of synaptic transmission from L30 inhibitory interneurons of* aplysia *depends on residual presynaptic Ca^{2+} but not on postsynaptic Ca^{2+}*, J. Neurophysiol., 78 (1997), pp. 2061–2071.

A. L. Fogelson and R. S. Zucker, *Presynaptic calcium diffusion from various arrays of single channels*, Biophys. J., 48 (1985), pp. 1003–1017.

M. Geppert, Y. Goda, R. E. Hammer, C. Li, T. W. Rosahl, C. F. Stevens, and T. C. Südhof, *Synaptotagmin I: A major Ca^{2+} sensor for transmitter release at a central synapse*, Cell, 79 (1994), pp. 717–727.

Y. Goda, *SNAREs and regulated vesicle exocytosis*, Proc. Natl. Acad. Sci. USA, 94 (1997), pp. 769–772.

R. Heidelberger, C. Heinemann, E. Neher, and G. Matthews, *Calcium dependence of the rate of exocytosis in a synaptic terminal*, Nature, 371 (1994), pp. 513–515.

B. Hille, *Modulation of ion-channel function by G-protein-coupled receptors*, Trends Neurosci., 17 (1994), pp. 531–536.

B. Hochner, H. Parnas, and I. Parnas, *Membrane depolarization evokes neurotransmitter release in the absence of calcium entry*, Nature, 342 (1989), pp. 433–435.

S. R. Ikeda, *Voltage-dependent modulation of N-type calcium channels by G-protein $\beta\gamma$ subunits*, Nature, 380 (1996), pp. 255–258.

S. E. Jarvis, W. Barr, Z. P. Feng, J. Hamid, and G. W. Zamponi, *Molecular determinants of syntaxin 1 modulation of N-type calcium channels*, J. Biol. Chem., 277 (2002), pp. 44399–44407.

H. Kamiya and R. S. Zucker, *Residual Ca^{2+} and short-term synaptic plasticity*, Nature, 371 (1994), pp. 603–606.

B. Katz and R. Miledi, *The role of calcium in neuromuscular facilitation*, J. Physiol. (Lond.), 195 (1968), pp. 481–492.

J. Klingauf and E. Neher, *Modeling buffered Ca^{2+} diffusion near the membrane: Implications for secretion in neuroendocrine cells*, Biophys. J., 72 (1997), pp. 674–690.

R. Llinás, I. Z. Steinberg, and K. Walton, *Presynaptic calcium currents and their relation to synaptic transmission: Voltage clamp study in squid giant synapse and theoretical model for the calcium gate*, Proc. Natl. Acad. Sci. USA, 73 (1976), pp. 2918–2922.

H. Markram, Y. Wang, and M. Tsodyks, *Differential signaling via the same axon of neocortical pyramidal neurons*, Proc. Natl. Acad. Sci. USA, 95 (1998), pp. 5323–5328.

V. Matveev, A. Sherman, and R. S. Zucker, *New and corrected simulations of synaptic function*, Biophys. J., 83 (2002), pp. 1368–1373.

V. Matveev, R. S. Zucker, and A. Sherman, *Facilitation through buffer saturation: constraints on endogenous buffering properties*, Biophys. J., 86 (2004), pp. 2691–2709.

M. Naraghi and E. Neher, *Linearized buffered Ca^{2+} diffusion in microdomains and its implications for calculation of $[Ca^{2+}]$ at the mouth of a calcium channel*, J. Neurosci., 17 (1997), pp. 6961–6973.

E. Neher, *Concentration profiles of intracellular calcium in the presence of a diffusible chelator*, in Calcium Electrogenesis and Neuronal Functioning, U. Heinemann, M. Klee, E. Neher, and W. Singer, eds., Springer-Verlag, Berlin, 1986, pp. 80–96.

———, *Usefulness and limitations of linear approximations to the understanding of Ca^{2+} signals*, Cell Calcium, 24 (1998), pp. 345–357.

H. Parnas, G. Hovav, and I. Parnas, *Effect of Ca^{2+} diffusion on the time course of neurotransmitter release*, Biophys. J., 55 (1989), pp. 859–874.

H. Parnas and L. A. Segel, *A theoretical study of calcium entry in nerve terminals, with application to neurotransmitter release*, J. theor. Biol., 91 (1981), pp. 125–169.

R. Pethig, M. Kuhn, E. Payne, T. Chen, and L. F. Jaffe, *On the dissociation constants of BAPTA-type Ca^{2+} buffers*, Cell Calcium, 10 (1989), pp. 491–498.

R. Rahamimoff, *A dual effect of calcium ions on neuromuscular facilitation*, J. Physiol. (Lond.), 195 (1968), pp. 471–480.

W. Rall, *Distinguishing theoretical synaptic potentials computed for different soma-dendritic distributions of synaptic inputs*, J. Neurophysiol., 30 (1967), pp. 1138–1168.

R. Rao-Mirotznik, A. B. Harkins, G. Buchsbaum, and P. Sterling, *Mammalian rod terminal: architecture of a binary synapse*, Neuron, 14 (1995), pp. 561–569.

J. Rettig and E. Neher, *Emerging roles of presynaptic proteins in Ca^{2+}-triggered exocytosis*, Science, 298 (2002), pp. 781–785.

W. M. Roberts, *Spatial calcium buffering in saccular hair cells*, Nature, 363 (1993), pp. 74–76.

R. Robitaille and M. P. Charlton, *Frequency facilitation is not caused by residual ionized calcium at the frog neuromuscular junction*, Ann. NY Acad. Sci., 635 (1991), pp. 492–494.

T. Schikorski and C. F. Stevens, *Quantitative ultrastructural analysis of hippocampal excitatory synapses*, J. Neurosci., 17 (1997), pp. 5858–5867.

W.-X. Shen and J. P. Horn, *Presynaptic muscarinic inhibition in bullfrog sympathetic ganglia*, J. Physiol. (Lond.), 491 (1996), pp. 413–421.

Z.-H. Sheng, J. Rettig, T. Cook, and W. A. Catterall, *Calcium-dependent interaction of N-type calcium channels with the synaptic core complex*, Nature, 379 (1996), pp. 451–454.

Z.-H. Sheng, C. T. Yokoyama, and W. A. Catterall, *Interaction of the synprint site of N-type Ca^{2+} channels with the C2B domain of synaptotagmin I*, Proc. Natl. Acad. Sci. USA, 94 (1997), pp. 5405–5410.

S. Simon and R. Llinás, *Compartmentalization of the submembrane calcium activity during calcium influx and its significance in transmitter release*, Biophys. J., 48 (1985), pp. 485–498.

G. D. Smith, *Analytical steady-state solution to the rapid buffering approximation near an open channel*, Biophys. J., 71 (1996), pp. 3064–3072.

G. D. Smith, L. Dai, R. M. Miura, and A. Sherman, *Asymptotic analysis of buffered calcium diffusion near a point source*, SIAM J. Appl. Math., 61 (2001), pp. 1816–1838.

G. D. Smith, J. E. Pearson, and J. E. Keizer, *Modeling intracellular calcium waves and sparks*, in Computational Cell Biology, C. P. Fall, E. S. Marland, J. M. Wagner, and J. J. Tyson, eds., Springer, New York, 2002, pp. 198–229.

E. F. Stanley, *Decline in calcium cooperativity as the basis of facilitation at the squid giant synapse*, J. Neurosci., 6 (1986), pp. 782–789.

———, *Single calcium channels and acetylcholine release at a presynaptic nerve terminal*, Neuron, 11 (1993), pp. 1007–1011.

T. C. Südhof, *The synaptic vesicle cycle: A cascade of protein-protein interactions*, Nature, 375 (1995), pp. 645–653.

Y. Tang, T. Schlumpberger, T. Kim, M. Lueker, and R. S. Zucker, *Effects of mobile buffers on facilitation: Experimental and computational studies*, Biophys. J., 78 (2000), pp. 2735–2751.

M. V. Tsodyks and H. Markram, *The neural code between neocortical pyramidal neurons depends on neurotransmitter release probability*, Proc. Natl. Acad. Sci. USA, 94 (1997), pp. 719–723.

J. Wagner and J. Keizer, *Effects of rapid buffer on Ca^{2+} diffusion and Ca^{2+} oscillations*, Biophys. J., 67 (1994), pp. 447–456.

J. L. Winslow, S. N. Duffy, and M. P. Charlton, *Homosynaptic facilitation of transmitter release in crayfish is not affected by mobile calcium chelators: implications for the residual ionized calcium hypothesis from electrophysiological and computational analyses*, J. Neurophysiol., 72 (1994), pp. 1769–1793.

L.-G. Wu and P. Saggau, *Presynaptic inhibition of elicited neurotransmitter release*, Trends Neurosci., 20 (1997), pp. 204–212.

W. M. Yamada and R. S. Zucker, *Time course of transmitter release calculated from simulations of a calcium diffusion model*, Biophys. J., 61 (1992), pp. 671–682.

G. W. Zamponi and T. P. Snutch, *Decay of prepulse facilitation of N type calcium channels during G protein inhibition is consistent with binding of a single $G_{\beta\gamma}$ subunit*, Proc. Natl. Acad. Sci. USA, 95 (1998), pp. 4035–4039.

R. S. Zucker, *Short-term synaptic plasticity*, Ann. Rev. Neurosci., 12 (1989), pp. 13–31.

R. S. Zucker and L. Landò, *Mechanism of transmitter release: voltage hypothesis and calcium hypothesis*, Science, 231 (1986), pp. 574–579.

R. S. Zucker and W. G. Regehr, *Short-term synaptic plasticity*, Annu. Rev. Physiol., 64 (2002), pp. 355–405.

R. S. Zucker and N. Stockbridge, *Presynaptic calcium diffusion and the time courses of transmitter release and synaptic facilitation at the squid giant synapse*, J. Neurosci., 3 (1983), pp. 1263–1269.

Lecture Notes in Mathematics

For information about Vols. 1–1673
please contact your bookseller or Springer

Vol. 1674: G. Klaas, C. R. Leedham-Green, W. Plesken, Linear Pro-p-Groups of Finite Width (1997)

Vol. 1675: J. E. Yukich, Probability Theory of Classical Euclidean Optimization Problems (1998)

Vol. 1676: P. Cembranos, J. Mendoza, Banach Spaces of Vector-Valued Functions (1997)

Vol. 1677: N. Proskurin, Cubic Metaplectic Forms and Theta Functions (1998)

Vol. 1678: O. Krupková, The Geometry of Ordinary Variational Equations (1997)

Vol. 1679: K.-G. Grosse-Erdmann, The Blocking Technique. Weighted Mean Operators and Hardy's Inequality (1998)

Vol. 1680: K.-Z. Li, F. Oort, Moduli of Supersingular Abelian Varieties (1998)

Vol. 1681: G. J. Wirsching, The Dynamical System Generated by the 3n+1 Function (1998)

Vol. 1682: H.-D. Alber, Materials with Memory (1998)

Vol. 1683: A. Pomp, The Boundary-Domain Integral Method for Elliptic Systems (1998)

Vol. 1684: C. A. Berenstein, P. F. Ebenfelt, S. G. Gindikin, S. Helgason, A. E. Tumanov, Integral Geometry, Radon Transforms and Complex Analysis. Firenze, 1996. Editors: E. Casadio Tarabusi, M. A. Picardello, G. Zampieri (1998)

Vol. 1685: S. König, A. Zimmermann, Derived Equivalences for Group Rings (1998)

Vol. 1686: J. Azéma, M. Émery, M. Ledoux, M. Yor (Eds.), Séminaire de Probabilités XXXII (1998)

Vol. 1687: F. Bornemann, Homogenization in Time of Singularly Perturbed Mechanical Systems (1998)

Vol. 1688: S. Assing, W. Schmidt, Continuous Strong Markov Processes in Dimension One (1998)

Vol. 1689: W. Fulton, P. Pragacz, Schubert Varieties and Degeneracy Loci (1998)

Vol. 1690: M. T. Barlow, D. Nualart, Lectures on Probability Theory and Statistics. Editor: P. Bernard (1998)

Vol. 1691: R. Bezrukavnikov, M. Finkelberg, V. Schechtman, Factorizable Sheaves and Quantum Groups (1998)

Vol. 1692: T. M. W. Eyre, Quantum Stochastic Calculus and Representations of Lie Superalgebras (1998)

Vol. 1694: A. Braides, Approximation of Free-Discontinuity Problems (1998)

Vol. 1695: D. J. Hartfiel, Markov Set-Chains (1998)

Vol. 1696: E. Bouscaren (Ed.): Model Theory and Algebraic Geometry (1998)

Vol. 1697: B. Cockburn, C. Johnson, C.-W. Shu, E. Tadmor, Advanced Numerical Approximation of Nonlinear Hyperbolic Equations. Cetraro, Italy, 1997. Editor: A. Quarteroni (1998)

Vol. 1698: M. Bhattacharjee, D. Macpherson, R. G. Möller, P. Neumann, Notes on Infinite Permutation Groups (1998)

Vol. 1699: A. Inoue,Tomita-Takesaki Theory in Algebras of Unbounded Operators (1998)

Vol. 1700: W. A. Woyczyński, Burgers-KPZ Turbulence (1998)

Vol. 1701: Ti-Jun Xiao, J. Liang, The Cauchy Problem of Higher Order Abstract Differential Equations (1998)

Vol. 1702: J. Ma, J. Yong, Forward-Backward Stochastic Differential Equations and Their Applications (1999)

Vol. 1703: R. M. Dudley, R. Norvaiša, Differentiability of Six Operators on Nonsmooth Functions and p-Variation (1999)

Vol. 1704: H. Tamanoi, Elliptic Genera and Vertex Operator Super-Algebras (1999)

Vol. 1705: I. Nikolaev, E. Zhuzhoma, Flows in 2-dimensional Manifolds (1999)

Vol. 1706: S. Yu. Pilyugin, Shadowing in Dynamical Systems (1999)

Vol. 1707: R. Pytlak, Numerical Methods for Optimal Control Problems with State Constraints (1999)

Vol. 1708: K. Zuo, Representations of Fundamental Groups of Algebraic Varieties (1999)

Vol. 1709: J. Azéma, M. Émery, M. Ledoux, M. Yor (Eds.), Séminaire de Probabilités XXXIII (1999)

Vol. 1710: M. Koecher, The Minnesota Notes on Jordan Algebras and Their Applications (1999)

Vol. 1711: W. Ricker, Operator Algebras Generated by Commuting Projections: A Vector Measure Approach (1999)

Vol. 1712: N. Schwartz, J. J. Madden, Semi-algebraic Function Rings and Reflectors of Partially Ordered Rings (1999)

Vol. 1713: F. Bethuel, G. Huisken, S. Müller, K. Steffen, Calculus of Variations and Geometric Evolution Problems. Cetraro, 1996. Editors: S. Hildebrandt, M. Struwe (1999)

Vol. 1714: O. Diekmann, R. Durrett, K. P. Hadeler, P. K. Maini, H. L. Smith, Mathematics Inspired by Biology. Martina Franca, 1997. Editors: V. Capasso, O. Diekmann (1999)

Vol. 1715: N. V. Krylov, M. Röckner, J. Zabczyk, Stochastic PDE's and Kolmogorov Equations in Infinite Dimensions. Cetraro, 1998. Editor: G. Da Prato (1999)

Vol. 1716: J. Coates, R. Greenberg, K. A. Ribet, K. Rubin, Arithmetic Theory of Elliptic Curves. Cetraro, 1997. Editor: C. Viola (1999)

Vol. 1717: J. Bertoin, F. Martinelli, Y. Peres, Lectures on Probability Theory and Statistics. Saint-Flour, 1997. Editor: P. Bernard (1999)

Vol. 1718: A. Eberle, Uniqueness and Non-Uniqueness of Semigroups Generated by Singular Diffusion Operators (1999)

Vol. 1719: K. R. Meyer, Periodic Solutions of the N-Body Problem (1999)

Vol. 1720: D. Elworthy, Y. Le Jan, X-M. Li, On the Geometry of Diffusion Operators and Stochastic Flows (1999)

Vol. 1721: A. Iarrobino, V. Kanev, Power Sums, Gorenstein Algebras, and Determinantal Loci (1999)

Vol. 1722: R. McCutcheon, Elemental Methods in Ergodic Ramsey Theory (1999)
Vol. 1723: J. P. Croisille, C. Lebeau, Diffraction by an Immersed Elastic Wedge (1999)
Vol. 1724: V. N. Kolokoltsov, Semiclassical Analysis for Diffusions and Stochastic Processes (2000)
Vol. 1725: D. A. Wolf-Gladrow, Lattice-Gas Cellular Automata and Lattice Boltzmann Models (2000)
Vol. 1726: V. Marić, Regular Variation and Differential Equations (2000)
Vol. 1727: P. Kravanja M. Van Barel, Computing the Zeros of Analytic Functions (2000)
Vol. 1728: K. Gatermann Computer Algebra Methods for Equivariant Dynamical Systems (2000)
Vol. 1729: J. Azéma, M. Émery, M. Ledoux, M. Yor (Eds.) Séminaire de Probabilités XXXIV (2000)
Vol. 1730: S. Graf, H. Luschgy, Foundations of Quantization for Probability Distributions (2000)
Vol. 1731: T. Hsu, Quilts: Central Extensions, Braid Actions, and Finite Groups (2000)
Vol. 1732: K. Keller, Invariant Factors, Julia Equivalences and the (Abstract) Mandelbrot Set (2000)
Vol. 1733: K. Ritter, Average-Case Analysis of Numerical Problems (2000)
Vol. 1734: M. Espedal, A. Fasano, A. Mikelić, Filtration in Porous Media and Industrial Applications. Cetraro 1998. Editor: A. Fasano. 2000.
Vol. 1735: D. Yafaev, Scattering Theory: Some Old and New Problems (2000)
Vol. 1736: B. O. Turesson, Nonlinear Potential Theory and Weighted Sobolev Spaces (2000)
Vol. 1737: S. Wakabayashi, Classical Microlocal Analysis in the Space of Hyperfunctions (2000)
Vol. 1738: M. Émery, A. Nemirovski, D. Voiculescu, Lectures on Probability Theory and Statistics (2000)
Vol. 1739: R. Burkard, P. Deuflhard, A. Jameson, J.-L. Lions, G. Strang, Computational Mathematics Driven by Industrial Problems. Martina Franca, 1999. Editors: V. Capasso, H. Engl, J. Periaux (2000)
Vol. 1740: B. Kawohl, O. Pironneau, L. Tartar, J.-P. Zolesio, Optimal Shape Design. Tróia, Portugal 1999. Editors: A. Cellina, A. Ornelas (2000)
Vol. 1741: E. Lombardi, Oscillatory Integrals and Phenomena Beyond all Algebraic Orders (2000)
Vol. 1742: A. Unterberger, Quantization and Non-holomorphic Modular Forms (2000)
Vol. 1743: L. Habermann, Riemannian Metrics of Constant Mass and Moduli Spaces of Conformal Structures (2000)
Vol. 1744: M. Kunze, Non-Smooth Dynamical Systems (2000)
Vol. 1745: V. D. Milman, G. Schechtman (Eds.), Geometric Aspects of Functional Analysis. Israel Seminar 1999-2000 (2000)
Vol. 1746: A. Degtyarev, I. Itenberg, V. Kharlamov, Real Enriques Surfaces (2000)
Vol. 1747: L. W. Christensen, Gorenstein Dimensions (2000)
Vol. 1748: M. Ruzicka, Electrorheological Fluids: Modeling and Mathematical Theory (2001)
Vol. 1749: M. Fuchs, G. Seregin, Variational Methods for Problems from Plasticity Theory and for Generalized Newtonian Fluids (2001)
Vol. 1750: B. Conrad, Grothendieck Duality and Base Change (2001)
Vol. 1751: N. J. Cutland, Loeb Measures in Practice: Recent Advances (2001)
Vol. 1752: Y. V. Nesterenko, P. Philippon, Introduction to Algebraic Independence Theory (2001)
Vol. 1753: A. I. Bobenko, U. Eitner, Painlevé Equations in the Differential Geometry of Surfaces (2001)
Vol. 1754: W. Bertram, The Geometry of Jordan and Lie Structures (2001)
Vol. 1755: J. Azéma, M. Émery, M. Ledoux, M. Yor (Eds.), Séminaire de Probabilités XXXV (2001)
Vol. 1756: P. E. Zhidkov, Korteweg de Vries and Nonlinear Schrödinger Equations: Qualitative Theory (2001)
Vol. 1757: R. R. Phelps, Lectures on Choquet's Theorem (2001)
Vol. 1758: N. Monod, Continuous Bounded Cohomology of Locally Compact Groups (2001)
Vol. 1759: Y. Abe, K. Kopfermann, Toroidal Groups (2001)
Vol. 1760: D. Filipović, Consistency Problems for Heath-Jarrow-Morton Interest Rate Models (2001)
Vol. 1761: C. Adelmann, The Decomposition of Primes in Torsion Point Fields (2001)
Vol. 1762: S. Cerrai, Second Order PDE's in Finite and Infinite Dimension (2001)
Vol. 1763: J.-L. Loday, A. Frabetti, F. Chapoton, F. Goichot, Dialgebras and Related Operads (2001)
Vol. 1764: A. Cannas da Silva, Lectures on Symplectic Geometry (2001)
Vol. 1765: T. Kerler, V. V. Lyubashenko, Non-Semisimple Topological Quantum Field Theories for 3-Manifolds with Corners (2001)
Vol. 1766: H. Hennion, L. Hervé, Limit Theorems for Markov Chains and Stochastic Properties of Dynamical Systems by Quasi-Compactness (2001)
Vol. 1767: J. Xiao, Holomorphic Q Classes (2001)
Vol. 1768: M.J. Pflaum, Analytic and Geometric Study of Stratified Spaces (2001)
Vol. 1769: M. Alberich-Carramiñana, Geometry of the Plane Cremona Maps (2002)
Vol. 1770: H. Gluesing-Luerssen, Linear Delay-Differential Systems with Commensurate Delays: An Algebraic Approach (2002)
Vol. 1771: M. Émery, M. Yor (Eds.), Séminaire de Probabilités 1967-1980. A Selection in Martingale Theory (2002)
Vol. 1772: F. Burstall, D. Ferus, K. Leschke, F. Pedit, U. Pinkall, Conformal Geometry of Surfaces in S^4 (2002)
Vol. 1773: Z. Arad, M. Muzychuk, Standard Integral Table Algebras Generated by a Non-real Element of Small Degree (2002)
Vol. 1774: V. Runde, Lectures on Amenability (2002)
Vol. 1775: W. H. Meeks, A. Ros, H. Rosenberg, The Global Theory of Minimal Surfaces in Flat Spaces. Martina Franca 1999. Editor: G. P. Pirola (2002)
Vol. 1776: K. Behrend, C. Gomez, V. Tarasov, G. Tian, Quantum Comohology. Cetraro 1997. Editors: P. de Bartolomeis, B. Dubrovin, C. Reina (2002)
Vol. 1777: E. García-Río, D. N. Kupeli, R. Vázquez-Lorenzo, Osserman Manifolds in Semi-Riemannian Geometry (2002)
Vol. 1778: H. Kiechle, Theory of K-Loops (2002)
Vol. 1779: I. Chueshov, Monotone Random Systems (2002)

Vol. 1780: J. H. Bruinier, Borcherds Products on O(2,1) and Chern Classes of Heegner Divisors (2002)
Vol. 1781: E. Bolthausen, E. Perkins, A. van der Vaart, Lectures on Probability Theory and Statistics. Ecole d' Eté de Probabilités de Saint-Flour XXIX-1999. Editor: P. Bernard (2002)
Vol. 1782: C.-H. Chu, A. T.-M. Lau, Harmonic Functions on Groups and Fourier Algebras (2002)
Vol. 1783: L. Grüne, Asymptotic Behavior of Dynamical and Control Systems under Perturbation and Discretization (2002)
Vol. 1784: L.H. Eliasson, S. B. Kuksin, S. Marmi, J.-C. Yoccoz, Dynamical Systems and Small Divisors. Cetraro, Italy 1998. Editors: S. Marmi, J.-C. Yoccoz (2002)
Vol. 1785: J. Arias de Reyna, Pointwise Convergence of Fourier Series (2002)
Vol. 1786: S. D. Cutkosky, Monomialization of Morphisms from 3-Folds to Surfaces (2002)
Vol. 1787: S. Caenepeel, G. Militaru, S. Zhu, Frobenius and Separable Functors for Generalized Module Categories and Nonlinear Equations (2002)
Vol. 1788: A. Vasil'ev, Moduli of Families of Curves for Conformal and Quasiconformal Mappings (2002)
Vol. 1789: Y. Sommerhäuser, Yetter-Drinfel'd Hopf algebras over groups of prime order (2002)
Vol. 1790: X. Zhan, Matrix Inequalities (2002)
Vol. 1791: M. Knebusch, D. Zhang, Manis Valuations and Prüfer Extensions I: A new Chapter in Commutative Algebra (2002)
Vol. 1792: D. D. Ang, R. Gorenflo, V. K. Le, D. D. Trong, Moment Theory and Some Inverse Problems in Potential Theory and Heat Conduction (2002)
Vol. 1793: J. Cortés Monforte, Geometric, Control and Numerical Aspects of Nonholonomic Systems (2002)
Vol. 1794: N. Pytheas Fogg, Substitution in Dynamics, Arithmetics and Combinatorics. Editors: V. Berthé, S. Ferenczi, C. Mauduit, A. Siegel (2002)
Vol. 1795: H. Li, Filtered-Graded Transfer in Using Noncommutative Gröbner Bases (2002)
Vol. 1796: J.M. Melenk, hp-Finite Element Methods for Singular Perturbations (2002)
Vol. 1797: B. Schmidt, Characters and Cyclotomic Fields in Finite Geometry (2002)
Vol. 1798: W.M. Oliva, Geometric Mechanics (2002)
Vol. 1799: H. Pajot, Analytic Capacity, Rectifiability, Menger Curvature and the Cauchy Integral (2002)
Vol. 1800: O. Gabber, L. Ramero, Almost Ring Theory (2003)
Vol. 1801: J. Azéma, M. Émery, M. Ledoux, M. Yor (Eds.), Séminaire de Probabilités XXXVI (2003)
Vol. 1802: V. Capasso, E. Merzbach, B.G. Ivanoff, M. Dozzi, R. Dalang, T. Mountford, Topics in Spatial Stochastic Processes. Martina Franca, Italy 2001. Editor: E. Merzbach (2003)
Vol. 1803: G. Dolzmann, Variational Methods for Crystalline Microstructure - Analysis and Computation (2003)
Vol. 1804: I. Cherednik, Ya. Markov, R. Howe, G. Lusztig, Iwahori-Hecke Algebras and their Representation Theory. Martina Franca, Italy 1999. Editors: V. Baldoni, D. Barbasch (2003)
Vol. 1805: F. Cao, Geometric Curve Evolution and Image Processing (2003)
Vol. 1806: H. Broer, I. Hoveijn. G. Lunther, G. Vegter, Bifurcations in Hamiltonian Systems. Computing Singularities by Gröbner Bases (2003)
Vol. 1807: V. D. Milman, G. Schechtman (Eds.), Geometric Aspects of Functional Analysis. Israel Seminar 2000-2002 (2003)
Vol. 1808: W. Schindler, Measures with Symmetry Properties (2003)
Vol. 1809: O. Steinbach, Stability Estimates for Hybrid Coupled Domain Decomposition Methods (2003)
Vol. 1810: J. Wengenroth, Derived Functors in Functional Analysis (2003)
Vol. 1811: J. Stevens, Deformations of Singularities (2003)
Vol. 1812: L. Ambrosio, K. Deckelnick, G. Dziuk, M. Mimura, V. A. Solonnikov, H. M. Soner, Mathematical Aspects of Evolving Interfaces. Madeira, Funchal, Portugal 2000. Editors: P. Colli, J. F. Rodrigues (2003)
Vol. 1813: L. Ambrosio, L. A. Caffarelli, Y. Brenier, G. Buttazzo, C. Villani, Optimal Transportation and its Applications. Martina Franca, Italy 2001. Editors: L. A. Caffarelli, S. Salsa (2003)
Vol. 1814: P. Bank, F. Baudoin, H. Föllmer, L.C.G. Rogers, M. Soner, N. Touzi, Paris-Princeton Lectures on Mathematical Finance 2002 (2003)
Vol. 1815: A. M. Vershik (Ed.), Asymptotic Combinatorics with Applications to Mathematical Physics. St. Petersburg, Russia 2001 (2003)
Vol. 1816: S. Albeverio, W. Schachermayer, M. Talagrand, Lectures on Probability Theory and Statistics. Ecole d'Eté de Probabilités de Saint-Flour XXX-2000. Editor: P. Bernard (2003)
Vol. 1817: E. Koelink, W. Van Assche(Eds.), Orthogonal Polynomials and Special Functions. Leuven 2002 (2003)
Vol. 1818: M. Bildhauer, Convex Variational Problems with Linear, nearly Linear and/or Anisotropic Growth Conditions (2003)
Vol. 1819: D. Masser, Yu. V. Nesterenko, H. P. Schlickewei, W. M. Schmidt, M. Waldschmidt, Diophantine Approximation. Cetraro, Italy 2000. Editors: F. Amoroso, U. Zannier (2003)
Vol. 1820: F. Hiai, H. Kosaki, Means of Hilbert Space Operators (2003)
Vol. 1821: S. Teufel, Adiabatic Perturbation Theory in Quantum Dynamics (2003)
Vol. 1822: S.-N. Chow, R. Conti, R. Johnson, J. Mallet-Paret, R. Nussbaum, Dynamical Systems. Cetraro, Italy 2000. Editors: J. W. Macki, P. Zecca (2003)
Vol. 1823: A. M. Anile, W. Allegretto, C. Ringhofer, Mathematical Problems in Semiconductor Physics. Cetraro, Italy 1998. Editor: A. M. Anile (2003)
Vol. 1824: J. A. Navarro González, J. B. Sancho de Salas, $\mathcal{C}^{\infty}$ - Differentiable Spaces (2003)
Vol. 1825: J. H. Bramble, A. Cohen, W. Dahmen, Multiscale Problems and Methods in Numerical Simulations, Martina Franca, Italy 2001. Editor: C. Canuto (2003)
Vol. 1826: K. Dohmen, Improved Bonferroni Inequalities via Abstract Tubes. Inequalities and Identities of Inclusion-Exclusion Type. VIII, 113 p, 2003.
Vol. 1827: K. M. Pilgrim, Combinations of Complex Dynamical Systems. IX, 118 p, 2003.
Vol. 1828: D. J. Green, Gröbner Bases and the Computation of Group Cohomology. XII, 138 p, 2003.

Vol. 1829: E. Altman, B. Gaujal, A. Hordijk, Discrete-Event Control of Stochastic Networks: Multimodularity and Regularity. XIV, 313 p, 2003.
Vol. 1830: M. I. Gil', Operator Functions and Localization of Spectra. XIV, 256 p, 2003.
Vol. 1831: A. Connes, J. Cuntz, E. Guentner, N. Higson, J. E. Kaminker, Noncommutative Geometry, Martina Franca, Italy 2002. Editors: S. Doplicher, L. Longo (2004)
Vol. 1832: J. Azéma, M. Émery, M. Ledoux, M. Yor (Eds.), Séminaire de Probabilités XXXVII (2003)
Vol. 1833: D.-Q. Jiang, M. Qian, M.-P. Qian, Mathematical Theory of Nonequilibrium Steady States. On the Frontier of Probability and Dynamical Systems. IX, 280 p, 2004.
Vol. 1834: Yo. Yomdin, G. Comte, Tame Geometry with Application in Smooth Analysis. VIII, 186 p, 2004.
Vol. 1835: O.T. Izhboldin, B. Kahn, N.A. Karpenko, A. Vishik, Geometric Methods in the Algebraic Theory of Quadratic Forms. Summer School, Lens, 2000. Editor: J.-P. Tignol (2004)
Vol. 1836: C. Năstăsescu, F. Van Oystaeyen, Methods of Graded Rings. XIII, 304 p, 2004.
Vol. 1837: S. Tavaré, O. Zeitouni, Lectures on Probability Theory and Statistics. Ecole d'Eté de Probabilités de Saint-Flour XXXI-2001. Editor: J. Picard (2004)
Vol. 1838: A.J. Ganesh, N.W. O'Connell, D.J. Wischik, Big Queues. XII, 254 p, 2004.
Vol. 1839: R. Gohm, Noncommutative Stationary Processes. VIII, 170 p, 2004.
Vol. 1840: B. Tsirelson, W. Werner, Lectures on Probability Theory and Statistics. Ecole d'Eté de Probabilités de Saint-Flour XXXII-2002. Editor: J. Picard (2004)
Vol. 1841: W. Reichel, Uniqueness Theorems for Variational Problems by the Method of Transformation Groups (2004)
Vol. 1842: T. Johnsen, A.L. Knutsen, K3 Projective Models in Scrolls (2004)
Vol. 1843: B. Jefferies, Spectral Properties of Noncommuting Operators (2004)
Vol. 1844: K.F. Siburg, The Principle of Least Action in Geometry and Dynamics (2004)
Vol. 1845: Min Ho Lee, Mixed Automorphic Forms, Torus Bundles, and Jacobi Forms (2004)
Vol. 1846: H. Ammari, H. Kang, Reconstruction of Small Inhomogeneities from Boundary Measurements (2004)
Vol. 1847: T.R. Bielecki, T. Björk, M. Jeanblanc, M. Rutkowski, J.A. Scheinkman, W. Xiong, Paris-Princeton Lectures on Mathematical Finance 2003 (2004)
Vol. 1848: M. Abate, J. E. Fornaess, X. Huang, J. P. Rosay, A. Tumanov, Real Methods in Complex and CR Geometry, Martina Franca, Italy 2002. Editors: D. Zaitsev, G. Zampieri (2004)
Vol. 1849: Martin L. Brown, Heegner Modules and Elliptic Curves (2004)
Vol. 1850: V. D. Milman, G. Schechtman (Eds.), Geometric Aspects of Functional Analysis. Israel Seminar 2002-2003 (2004)
Vol. 1851: O. Catoni, Statistical Learning Theory and Stochastic Optimization (2004)
Vol. 1852: A.S. Kechris, B.D. Miller, Topics in Orbit Equivalence (2004)
Vol. 1853: Ch. Favre, M. Jonsson, The Valuative Tree (2004)
Vol. 1854: O. Saeki, Topology of Singular Fibers of Differential Maps (2004)
Vol. 1855: G. Da Prato, P.C. Kunstmann, I. Lasiecka, A. Lunardi, R. Schnaubelt, L. Weis, Functional Analytic Methods for Evolution Equations. Editors: M. Iannelli, R. Nagel, S. Piazzera (2004)
Vol. 1856: K. Back, T.R. Bielecki, C. Hipp, S. Peng, W. Schachermayer, Stochastic Methods in Finance, Bressanone/Brixen, Italy, 2003. Editors: M. Fritelli, W. Runggaldier (2004)
Vol. 1857: M. Émery, M. Ledoux, M. Yor (Eds.), Séminaire de Probabilités XXXVIII (2005)
Vol. 1858: A.S. Cherny, H.-J. Engelbert, Singular Stochastic Differential Equations (2005)
Vol. 1859: E. Letellier, Fourier Transforms of Invariant Functions on Finite Reductive Lie Algebras (2005)
Vol. 1860: A. Borisyuk, G.B. Ermentrout, A. Friedman, D. Terman, Tutorials in Mathematical Biosciences I. Mathematical Neurosciences (2005)
Vol. 1861: G. Benettin, J. Henrard, S. Kuksin, Hamiltonian Dynamics - Theory and Applications, Cetraro, Italy, 1999. Editor: A. Giorgilli (2005)
Vol. 1862: B. Helffer, F. Nier, Hypoelliptic Estimates and Spectral Theory for Fokker-Planck Operators and Witten Laplacians (2005)
Vol. 1863: H. Führ, Abstract Harmonic Analysis of Continuous Wavelet Transforms (2005)
Vol. 1864: K. Efstathiou, Metamorphoses of Hamiltonian Systems with Symmetries (2005)
Vol. 1865: D. Applebaum, B.V. R. Bhat, J. Kustermans, J. M. Lindsay, Quantum Independent Increment Processes I. From Classical Probability to Quantum Stochastic Calculus. Editors: M. Schürmann, U. Franz (2005)
Vol. 1866: O.E. Barndorff-Nielsen, U. Franz, R. Gohm, B. Kümmerer, S. Thorbjønsen, Quantum Independent Increment Processes II. Structure of Quantum Levy Processes, Classical Probability, and Physics. Editors: M. Schürmann, U. Franz, (2005)
Vol. 1867: J. Sneyd (Ed.), Tutorials in Mathematical Biosciences II. Mathematical Modeling of Calcium Dynamics and Signal Transduction. (2005)
Vol. 1868: J. Jorgenson, S. Lang, $Pos_n(R)$ and Eisenstein Sereies. (2005)
Vol. 1869: A. Dembo, T. Funaki, Lectures on Probability Theory and Statistics. Ecole d'Eté de Probabilités de Saint-Flour XXXIII-2003. Editor: J. Picard (2005)

Recent Reprints and New Editions

Vol. 1200: V. D. Milman, G. Schechtman (Eds.), Asymptotic Theory of Finite Dimensional Normed Spaces. 1986. – Corrected Second Printing (2001)
Vol. 1471: M. Courtieu, A.A. Panchishkin, Non-Archimedean L-Functions and Arithmetical Siegel Modular Forms. – Second Edition (2003)
Vol. 1618: G. Pisier, Similarity Problems and Completely Bounded Maps. 1995 – Second, Expanded Edition (2001)
Vol. 1629: J.D. Moore, Lectures on Seiberg-Witten Invariants. 1997 – Second Edition (2001)
Vol. 1638: P. Vanhaecke, Integrable Systems in the realm of Algebraic Geometry. 1996 – Second Edition (2001)

Vol. 1702: J. Ma, J. Yong, Forward-Backward Stochastic Differential Equations and their Applications. 1999. – Corrected 3rd printing (2005)

Printing: Krips bv, Meppel
Binding: Stürtz, Würzburg